State Transportation Liaison Funded Positions Study

October 2009

U.S. Department of Transportation
Federal Highway Administration
Office of Project Development and Environmental Review

Prepared with assistance from:
U.S. Department of Transportation
Research and Innovative Technology Administration
John A. Volpe National Transportation Systems Center

Acknowledgments

This U.S. Department of Transportation (USDOT) Federal Highway Administration (FHWA) report was prepared with assistance from the John A. Volpe National Transportation Systems Center (Volpe Center). The project team included David Carlson of FHWA's Office of Project Development and Environmental Review; Sharon Chan Edmiston, Alisa Zlotoff, and Alexandra Miller of the Volpe Center's Transportation Policy, Planning and Organizational Excellence Division; and Gina Filosa of Cambridge Systematics.

The project team wishes to thank the numerous individuals from State DOTs and resource agencies who graciously offered their time, knowledge, and assistance in the development of this study.

Table of Contents

Acknowledgments i
1 Introduction 1
    1.1 Background 1
    1.2 Study Objectives and Report Organization 1
    1.3 Study Methodology 2
2 General Findings 8
    2.1 Program Attributes 8
    2.2 Overview: The Decisionmaking Process 13
3 Stage 1: Assessing the Need and Demand for Funded Positions 17
    3.1 Conducting a Baseline Assessment 17
    3.2 Justifying and Quantifying the Need 20
4 Stage 2: Generating Program Support 22
    4.1 Gaining Support from Management and Government Officials 22
    4.2 Implementing a Pilot Program 23
5 Stage 3: Designing a Funded Positions Program 24
    5.1 Locating the Funded Positions 25
    5.2 Determining Term Lengths 25
    5.3 Determining Experience and Grade Levels 26
    5.4 Examining Centralized and Decentralized Management 26
    5.5 Providing Training Opportunities 26
6 Stage 4: Formalizing Interagency Agreements 29
    6.1 Modeling Agreements on Existing Agreements 29
    6.2 Renegotiating Agreements 31
7 Stage 5: Implementing and Managing the Program 33
    7.1 Finding and Hiring Liaisons 34
    7.2 Resolving Institutional/Interagency Relationship Issues 37
    7.3 Involving Liaisons in SAFETEA-LU Planning Activities 39
8 Stage 6: Evaluating Program Outcomes 43
    8.1 Establishing Performance Measures 43
    8.2 Quantitative versus Qualitative Performance Measures 44
    8.3 Examples of Existing Performance Measures and Tools for Measurement ... 45
    8.4 Use of Performance Measures 48
    8.5 Measurement of Streamlining of Benefits 49
    8.6 Administrative Reporting and Providing Feedback 49
    8.7 Performance Appraisals 51

APPENDIX A: Interview List.. **53**
APPENDIX B: Selected Bibliography **55**
APPENDIX C: Discussion Guides... **63**
APPENDIX D: Funded Positions Agreements.. **74**

List of Figures

Figure 1. Funded Positions Program Quadrant.. 8
Figure 2. Six-Stage Decisionmaking Process 13
Figure 3: Assessing the Need and Demand for Funded Positions 17
Figure 4. Generating Program Support 22
Figure 5. Designing a Funded Positions Program 24
Figure 6. Formalizing Interagency Agreements 29
Figure 7. Implementing and Managing the Program 33
Figure 8. Evaluating Program Outcomes 43

List of Tables

Table 1. Summary of General Reports on Funded Positions*... 3
Table 2. Summary of Interagency Agreements* 4
Table 3. Funded Positions in Federal Agencies, by State*.. 9
Table 4. Funded Positions in State Agencies, by State 10
Table 5. Duration of Service for Funded Positions... 25
Table 6. Number of Positions Funded per Agreement 30
Table 7. Funding Amounts per Agreement.. 30
Table 8. Funding Sources for Funded Positions 30
Table 9. Performance Measurement and Reporting Tools* 47

Acronyms and Abbreviations

AASHTO	American Association of State Highway and Transportation Officials
ADOT	Arizona Department of Transportation
AHTD	Arkansas Highway Transportation Department
Caltrans	California Department of Transportation
CDOT	Colorado Department of Transportation
CSS	Context Sensitive Solutions
CWA	Clean Water Act
DOT	Department of Transportation
ETDM	Efficient Transportation Decision-Making
EPA	Environmental Protection Agency
ESA	Endangered Species Act
FDACS	Florida Department of Agriculture and Consumer Services
FHWA	Federal Highway Administration
FTE	full-time employee
FWCC	Fish and Wildlife Conservation Commission
FWP	Fish, Wildlife and Parks
FWS	U.S. Fish and Wildlife Service
GIS	global information systems
INDOT	Indiana Department of Transportation
IPA	Intergovernmental Personnel Act
ITD	Idaho Transportation Department
KYTC	Kentucky Transportation Cabinet
LA DOTD	Louisiana Department of Transportation and Development
MaineDOT	Maine Department of Transportation
MDT	Montana Department of Transportation
MOA	Memorandum of Agreement
MOU	Memorandum of Understanding
MPO	Metropolitan Planning Organization
NCDENR	North Carolina Department of Environment and Natural Resources
NCDOT	North Carolina Department of Transportation
NCHRP	National Cooperative Highway Research Program
NEPA	National Environmental Policy Act
NHPA	National Historic Preservation Act
NJDOT	New Jersey Department of Transportation
NMFS	National Marine Fisheries Services
NOAA	National Oceanic and Atmospheric Administration
NOC	Notice of Coverage
NYSDOT	New York State Department of Transportation
PEL	Planning and Environment Linkages (FHWA)
PennDOT	Pennsylvania Department of Transportation

SAFETEA-LU	Safe, Accountable, Flexible, Efficient Transportation Equity Act: A Legacy for Users
SCDOT	South Carolina Department of Transportation
SHPO	State Historic Preservation Office
State DOT	State Department of Transportation
TDOT	Tennessee Department of Transportation
TEA-21	Transportation Equity Act for the 21st Century
TxDOT	Texas Department of Transportation
UDOT	Utah Department of Transportation
USACE	U.S. Army Corps of Engineers
USCG	U.S. Coast Guard
USFS	U.S. Forest Service
USFWS	U.S. Fish and Wildlife Service
Volpe Center	Volpe National Transportation Systems Center
WisDOT	Wisconsin Department of Transportation
WSDOT	Washington State Department of Transportation

Executive Summary

The Safe, Accountable, Flexible, Efficient Transportation Equity Act: A Legacy for Users (SAFETEA-LU), which was signed into law in August 2005, contained several provisions focused on streamlining the environmental review process. One of these provisions, Section 6002, allowed for State Department of Transportation (State DOT) funding of staff, at both Federal and State resource agencies, who are dedicated to working on State DOT projects on environmental streamlining and related planning activities.

This report assesses trends in the use of these "funded positions" and provides recommendations to State DOTs and resource agencies to support more effective uses of funded positions. The report is based on a study conducted by the Federal Highway Administration (FHWA) Office of Project Development and Environmental Review with assistance from the Volpe National Transportation Systems Center (Volpe Center). The study consisted of two parts: (1) a literature review to assess the state of the knowledge about State DOT-funded positions and agreements, and (2) a series of interviews with participants in funded positions programs, including program managers at State DOTs and Federal and State resource agencies and individuals in those positions.

Key findings from the study were:

- **Assessing the need for funded positions.** Agencies used funded positions as a mechanism to help address a variety of challenges, including difficulties related to project delivery, a need to improve communication or dialogue among agencies, and a need to better link planning and environmental review processes.

- **Formalizing funding agreements.** Developing a Memorandum of Understanding (MOU) or other agreement to formalize the funded positions program helped liaisons, State DOTs, and resource agencies to define roles and responsibilities. Agreements typically outlined program objectives and performance measures, provided an overview of the agency's mission, and detailed roles and responsibilities for liaisons, State DOTs, and resource agencies.

- **Finding and hiring liaisons.** Funded positions require strong written and oral communication skills, a clear understanding of the agency's mission and goals, and the ability to address sometimes competing sets of demands. Previous experience in transportation planning, community development, and/or conflict resolution can also be valuable assets for a funded position. Hiring-related challenges included recruiting qualified candidates for short-term liaison positions. Many States found that a five-year funding term helped to minimize staff turnover.

- **Providing training opportunities.** Funded positions reported that access to appropriate training provided them with the ability to more effectively navigate the permitting process. Agencies developed a variety of training opportunities for funded positions, including:
    - Orientation training for all new funded positions.
    - A "mentor" system for new liaisons to work closely with existing liaisons.

- Networking opportunities for liaisons to share information.
- External conferences or courses at training centers.

- **Resolving institutional/interagency relationship issues.** Most interviewees agreed that liaisons play an essential role in establishing priorities, mediating conflict, and encouraging strong working relationships between State DOTs and resource agencies. Many agencies emphasized the importance of clearly communicating State DOT project priorities to funded liaisons through regularly scheduled meetings or telephone calls. Liaisons with a single point of contact at the State DOT often found it easier to negotiate effectively between the State DOT and the resource agency. Having this single point of contact helped to establish a strong advocate for liaisons at the State DOT, facilitated communication of project priorities to liaisons, and provided oversight of liaisons.

- **Involving liaisons in planning activities.** Several resource agencies and State DOTs encouraged or in some cases required the involvement of funded positions in transportation planning activities, such as commenting on the regional transportation plans of metropolitan planning organizations (MPOs). Most funded positions agreed that involvement in long-range planning activities with the State DOT and local planning agencies such as MPOs helped lead to better decisionmaking during project development or to the resolution of conflicts early on. Some State DOTs reported a concern that involving liaisons in long-range planning would preclude them from focusing on more immediate permitting requests. Despite this concern, allowing liaisons to devote some time to planning activities often resulted in significant short- and long-term benefits, including:
  - Facilitation of better communication between the resource agency and local planning agencies during project planning stages.
  - More effective communication of resource agency objectives to MPOs.
  - Increased trust and stronger working relationships among State DOTs, resource agencies, and local planning agencies.
  - More integrated planning activities, which can lead to a seamless decisionmaking process that minimizes duplication of effort, promotes environmental stewardship, and reduces delays in project implementation.

- **Establishing performance measures.** Agencies in the early stages of implementing a funded positions program tended to utilize quantitative evaluation metrics, such as permit-turnaround time and number of permits approved, to evaluate liaisons' performance. More mature programs had also integrated qualitative metrics into performance evaluation. Examples of qualitative metrics included liaisons' ability to mediate conflict, provide high-quality comments on State DOT projects, and foster strong relationships between State DOTs and resource agencies.

This report concludes by outlining the stages of developing and implementing a funded positions program and describing key decisions at each stage.

1 Introduction

1.1 Background

Conducting an efficient and effective environmental review process requires coordination and collaboration between State transportation agencies and resource agencies. However, numerous demands on resource agency staff time often limit the ability of staff to participate in State transportation project planning or to expedite project reviews. Section 1309 of the Transportation Equity Act for the 21st Century (TEA-21), which was enacted in 1998, mandated an environmental streamlining process that improved transportation project delivery while protecting and enhancing the environment. One of the key elements of this process was cooperation between transportation and environmental resource agencies to develop and adhere to realistic project-development timeframes. Recognizing insufficient staff levels as a barrier to streamlining the environmental review process, TEA-21 allowed States to use Federal-aid project funds to provide additional resources to agencies that participate in the process, including Federal and State agencies and federally recognized Indian Tribes.

Building on and expanding the TEA-21 foundation, SAFETEA-LU was signed into law in August 2006. Several SAFETEA-LU provisions focused on improving efficiency in the highway program and project delivery. The Act also maintained Federal-aid project funds to support expedited environmental review and expanded eligibility of funding to include transportation planning activities. As outlined in SAFETEA-LU Section 6002 and codified in 23 USC Section 139(j), activities for which funds may be provided include transportation planning activities that precede the initiation of the environmental review process, dedicated staffing, training of agency personnel, information-gathering and mapping, and development of programmatic agreements.

In many cases, such funds have been used to employ staff at resource agencies who are dedicated to working on State DOT projects. The terms "funded positions," "external liaisons," and "funded liaisons" refer to dedicated staff (commonly housed at regulatory or resource agencies) funded by State DOTs to work on matters such as expedited project review and delivery. While many States have chosen to utilize federal funds to provide staff support to resource agencies, several, including California, North Carolina, and Washington, have opted to finance such support with State funds. Use of State transportation funds in this manner typically requires approval by the State legislature.

The number of funded positions has increased dramatically in the last decade, both in terms of the total number of positions and the number of States funding these positions. In 2003, there were 222 positions in 18 States, while in 2005 there were 375 funded positions in 34 States.[1] Although funded positions are becoming more widespread, they remain a relatively new policy implementation, and State DOTs and resource agencies are constantly making progress in understanding and using these positions.

1.2 Study Objectives and Report Organization

The FHWA Office of Project Development and Environmental Review has an interest in monitoring and assessing trends in the use of funded positions, as well as in providing technical assistance to State DOTs and resource agencies to support a more effective use of funded

---

[1] *State DOT-Funded Positions and Other Support to Resource and Regulatory Agencies, Tribes, and Non-Governmental Organizations for Environmental Stewardship and Streamlining Initiatives.* AASHTO Center for Environmental Excellence, May 2005.

positions in environmental streamlining activities. As part of this effort, FHWA is interested in identifying the following information:

- Current trends in the use of SAFETEA-LU Section 6002(j) funded positions.
- Best practices for establishing and maintaining State DOT-funded positions.
- Specific challenges and needs in order to develop technical assistance tools.
- An understanding of specific issues, such as how funded position liaisons can support the SAFETEA-LU provisions, which link planning and environmental review activities.

This report begins with an outline of the background, objectives, and study methodology, and proceeds with an overview of general findings. The remainder of the report is organized in six "stages" that mirror the process of developing a funded positions program from the ground up. This procedural and chronological approach is helpful because State DOTs and resource agencies have widely varying levels of experience with funded positions. For States that have not yet instituted funded positions programs, the staged approach provides a helpful guide to creating a funded positions program from beginning to end. Meanwhile, representatives from agencies with extensive experience with funded positions can find information relevant to their management and organizational questions without sorting through the more elementary information about why and how to establish a program.

The six stages of developing and managing a funded positions program are as follows:

- **Stage 1:** Assessing the need and demand for funded positions
- **Stage 2:** Generating program support
- **Stage 3:** Designing a funded positions program
- **Stage 4:** Formalizing interagency agreements
- **Stage 5:** Implementing and managing the program
- **Stage 6:** Evaluating program outcomes

1.3 Study Methodology

This study proceeded in two parts: (1) a literature review to assess the state of the knowledge about State DOT-funded positions and agreements, and (2) a series of interviews of State DOT and resource agency personnel on the use of funded positions.

1.3.1 Review of General Literature

The literature review, conducted in early 2008, evaluated general literature on funded positions as well as interagency, master, and operating agreements; template interagency agreements; funded positions program manuals; and performance tracking tools. The general literature on funded positions included six primary documents, which are summarized in Table 1. A summary of the interagency agreements that were reviewed is found in Table 2.

Table 1. Summary of General Reports on Funded Positions*

Report Reviewed	Relevant Topics	Agencies Covered
State DOT-Funded Positions and Other Support to Resource and Regulatory Agencies, Tribes, and Non-Governmental Organizations for Environmental Stewardship and Streamlining Initiatives. AASHTO Center for Environmental Excellence, May 2005.	<ul><li>Benefits of external positions</li><li>Number of funded positions nationally and at agencies</li><li>Trends in State DOT-funded positions</li><li>Other methods of State DOT-funded external support (e.g., GIS mapping, partnership efforts)</li><li>Administrative aspects of funding arrangements</li><li>Challenges of funded positions</li><li>Performance measures of funded positions</li><li>Training for funded positions</li><li>Lessons learned</li></ul>	<ul><li>State DOTs in all 50 States, Washington, D.C., and Puerto Rico</li></ul>
Implementing Performance Measurement in Environmental Streamlining. FHWA, May 2007.	<ul><li>Problems with streamlining efforts</li><li>Relationships</li><li>Communications</li><li>Timeliness</li><li>Performance</li><li>General issues and concerns</li></ul>	<ul><li>National Gallup Poll survey of transportation and resource agencies</li></ul>
"Measuring Environmental Performance at State Transportation Agencies." Marie Venner, *Transportation Research Record*, Paper No. 03-4485, pp. 9–18.	<ul><li>Individual performance accountability and funded positions</li></ul>	<ul><li>Caltrans</li><li>CDOT</li><li>NJDOT</li><li>NYSDOT</li><li>NCDOT</li><li>PennDOT</li><li>SCDOT</li><li>WSDOT</li></ul>
Meeting 2: Work Assignments and Performance in External Positions Funded by State DOTs. NCHRP Project 08-36, Peer Exchange Series on State and Metropolitan Transportation Planning Issues, December 2007 (unpublished).	<ul><li>Types of external positions</li><li>Management of external positions</li><li>Lessons learned</li><li>External position descriptions</li><li>Agreements used to fund external positions</li><li>Benefits of external positions</li><li>Issues and challenges related to external positions</li><li>Opportunities to improve external positions</li><li>Evaluation of external position performance</li><li>External positions and the planning process</li></ul>	<ul><li>Caltrans</li><li>FHWA Headquarters</li><li>FDOT</li><li>Georgia Department of Natural Resources</li><li>North Carolina Department of Environment and Natural Resources</li><li>NCDOT</li><li>Ohio DOT</li><li>Oregon DOT</li><li>PennDOT</li><li>Pennsylvania Fish and Boat Commission</li><li>SCDOT</li><li>USACE</li><li>USFWS</li><li>WisDOT</li></ul>
State DOT Positions at Resource Agencies: Distribution, Limitations, Accomplishments, and Maintaining Accountability. Report by AASHTO's	<ul><li>Number of funded positions at State DOTs</li><li>Performance measures</li><li>Managing staff and projects</li><li>Benefits</li></ul>	<ul><li>State DOTs in all 50 states, Washington, D.C., and Puerto Rico</li></ul>

Report Reviewed	Relevant Topics	Agencies Covered
Environmental Technical Assistance Program, August 2001.	• Results superseded by AASHTO's May 2005 report	

*DOT = Department of Transportation, AASHTO = American Association of State Highway and Transportation Officials, GIS = global information systems, FHWA = Federal Highway Administration, Caltrans = California Department of Transportation, CDOT = Colorado Department of Transportation, NJDOT = New Jersey Department of Transportation, NYSDOT = New York State Department of Transportation, NCDOT = North Carolina Department of Transportation, PennDOT = Pennsylvania Department of Transportation, SCDOT = South Carolina Department of Transportation, WSDOT = Washington State Department of Transportation, FDOT = Florida Department of Transportation, USACE = U.S. Army Corps of Engineers, USFWS = U.S. Fish and Wildlife Service, WisDOT = Wisconsin Department of Transportation.

Table 2. Summary of Interagency Agreements*

Funding Agency	No. of Agreements Reviewed	Agreement Mechanisms	Agencies Funded
Arkansas Highway and Transportation Department (with FHWA)	3	• MOA (1) • Cooperative Agreement (2)	• Department of Arkansas Heritage • USFWS • USACE
ADOT	2	• Letter of Intent • Letter of Intent and Operating Agreement	• Arizona SHPO • USACE
Caltrans	1	Standard Agreement	USFWS
FDOT	18	Funding Agreements	• Florida Department of Agriculture and Consumer Services • Florida Department of Community Affairs • Department of Environmental Protection • National Marine Fisheries Services • National Park Service • South Florida Water Management District • St. Johns River Water Management District • Suwanee River Water Management District • Southwest Florida Water Management District • Northwest Florida Water Management District • USACE • USEPA • USFS • USFWS • Florida Fish and Wildlife Conservation Commission • Florida SHPO
IDT	1	Cooperative Agreement	Idaho SHPO
INDOT	1	MOU	Indiana Department of Natural Resources
Kentucky	1	Cooperative Agreement	USFWS

Funding Agency	No. of Agreements Reviewed	Agreement Mechanisms	Agencies Funded
Transportation Cabinet			
LA DOTD (with FHWA)	2	• Intergovernmental Agreement • Interagency Agreement	• USFWS • Louisiana Department of Culture, Recreation and Tourism, Office of Cultural Development, Division of Archaeology
MaineDOT	2	Service Agreements	• Maine Department of Conservation Natural Areas Program • Maine Historic Preservation Commission
Mississippi Transportation Commission	1	MOA	USFWS
MDT (and FHWA)	2	Cooperative Agreement	• USFWS • Montana Department of Fish, Wildlife and Parks
NJDOT	1	MOA	• NJDEP HPO
PennDOT	5	• Cooperative Agreement (2) • MOU (3)	• USEPA • USFWS • Pennsylvania Department of Agriculture • Pennsylvania Game Commission • Pennsylvania Historical and Museum Commission
TDOT	1	Agreement	• Tennessee Department of Environment and Conservation
FHWA (at request of TxDOT)	3	• "Texas State law prohibits TxDOT from funding Federal agencies directly … the FHWA, at the request of TxDOT, has agreed to withhold funds from TxDOT's allocated discretionary funds."[†] • Interagency Agreement between FHWA and agencies	• USACE • USEPA • USFWS
WSDOT	1	State Interagency Agreement	Washington State Department of Ecology
Total	45		

*ADOT = Arizona Department of Transportation, FHWA = Federal Highway Administration, MOA = Memorandum of Agreement, USFWS = U.S. Fish and Wildlife Service, USACE = U.S. Army Corps of Engineers, SHPO = State Historic Preservation Office, FDOT = Florida DOT, EPA = Environmental Protection Agency, USFS = U.S. Forest Service, ITD = Idaho Transportation Department, INDOT = Indiana DOT, LA DOTD = Louisiana Department of Transportation and Development, MTDOT = Montana DOT, NJDOT = New Jersey DOT, PennDOT = Pennsylvania DOT, MOU = Memorandum of Understanding, TDOT = Tennessee DOT, TxDOT = Texas DOT, WSDOT = Washington State DOT.
[†] Source: TxDOT and USEPA Interagency Agreement, October 29, 2003.

In 2001, the American Association of State Highway and Transportation Officials' (AASHTO) Environmental Technical Assistance Team[2] produced a report that detailed the number of funded positions at State DOTs and described performance measures, management issues, and program benefits. Following up on this study, the AASHTO Center for Environmental Excellence sponsored an additional survey and interviews in 2005 with all 50 State DOTs.[3] The effort focused on State DOT-funded positions with resource agencies and Tribes. Information was also collected on other types of external support and partnerships funded by State DOTs.

In 2003, Venner Research and Consulting reported on transportation agencies' efforts to measure environmental performance and provided detailed case studies from several State DOTs.[4] The document detailed some challenges to developing performance measures, including those used to assess funded positions, such as skepticism from staff and management about the benefits of benchmarking and tracking work tasks. On behalf of FHWA, the Gallup Organization conducted a baseline study in 2003[5] to better understand key issues facing resource and transportation agencies in streamlining their environmental review processes. Gallup conducted a second study in 2006 to assess the changes in environmental streamlining practices that had occurred since the 2003 study.[6]

In 2007, the National Cooperative Highway Research Program (NCHRP) held a peer exchange that explored analysis of work assignments and performance in State DOT-funded external positions. Proceedings of the peer exchange documented participants' experiences with externally funded positions as well as lessons learned, program benefits, and general issues and challenges related to the positions.[7]

Although the existing literature provided information on critical issues related to State DOT-funded positions, it did not sufficiently address several areas, including:

- Whether funded positions support SAFETEA-LU provisions linking planning and the National Environmental Policy Act (NEPA) review process.

- Whether States with mature funded positions programs have developed more detailed programmatic agreements for liaison positions.

- Whether a resource agency receiving funded positions links the work of the positions to its strategic goals.

- How resource agencies or State DOTs can assess the need for a funded positions program or evaluate the need to expand, contract, or terminate an existing program.

- How resource agencies or State DOTs can justify or communicate the need for a funded positions program to upper management or financial staff.

- How resource agencies or State DOTs can develop appropriate performance measures to evaluate the success of funded positions in meeting program goals and objectives.

---

[2] *State DOT Positions at Resource Agencies: Distribution, Limitations, Accomplishments, and Maintaining Accountability.* Report by AASHTO's Environmental Technical Assistance Program, August 2001.

[3] *State DOT-Funded Positions and Other Support to Resource and Regulatory Agencies, Tribes, and Non-Governmental Organizations for Environmental Stewardship and Streamlining Initiatives.* AASHTO Center for Environmental Excellence, May 2005.

[4] Marie Venner, *Transportation Research Record*, Paper No. 03-4485, pp. 9–18, 2003.

[5] *Implementing Performance Measurement in Environmental Streamlining.* FHWA, August 2003.

[6] *Final Report: Implementing Performance Measurement in Environmental Streamlining.* FHWA, May 2007.

[7] *Meeting 2: Work Assignments and Performance in External Positions Funded by State DOTs.* NCHRP Project 08-36, Peer Exchange Series on State and Metropolitan Transportation Planning Issues, December 2007 (unpublished).

1.3.2 Interviews

The study interviews sought to address gaps in the existing literature by identifying current trends in funded positions programs and outlining decisionmaking processes during each stage of developing and implementing a program. Thirty-five telephone interviews were conducted over a ten-month period (July 2008 through April 2009) with participants in funded positions programs, including program managers at State DOTs and resource agencies and individuals in the funded positions (see Appendix A).

2 General Findings

2.1 Program Attributes

To examine a diversity of program experiences, eight States — California, Florida, North Carolina, South Carolina, Ohio, Tennessee, Utah, and Washington — were chosen to participate in the study interviews. The funded positions programs in these States varied in size (i.e., number of liaisons funded at one or more resource agencies) and maturity (i.e., number of years of program experience) (see Figure 1 and Table 3). States with larger programs were California (31 liaisons), North Carolina (27), Washington (25), and Florida (23); those with smaller programs were Ohio (five liaisons), Tennessee (six), and South Carolina (six). Utah had previously supported one funded position but terminated its program in 2009. Program size and maturity were corresponding factors: States with larger programs generally had more years of program experience than did States with smaller programs. Table 4 provides the number of funded positions in State agencies by State.

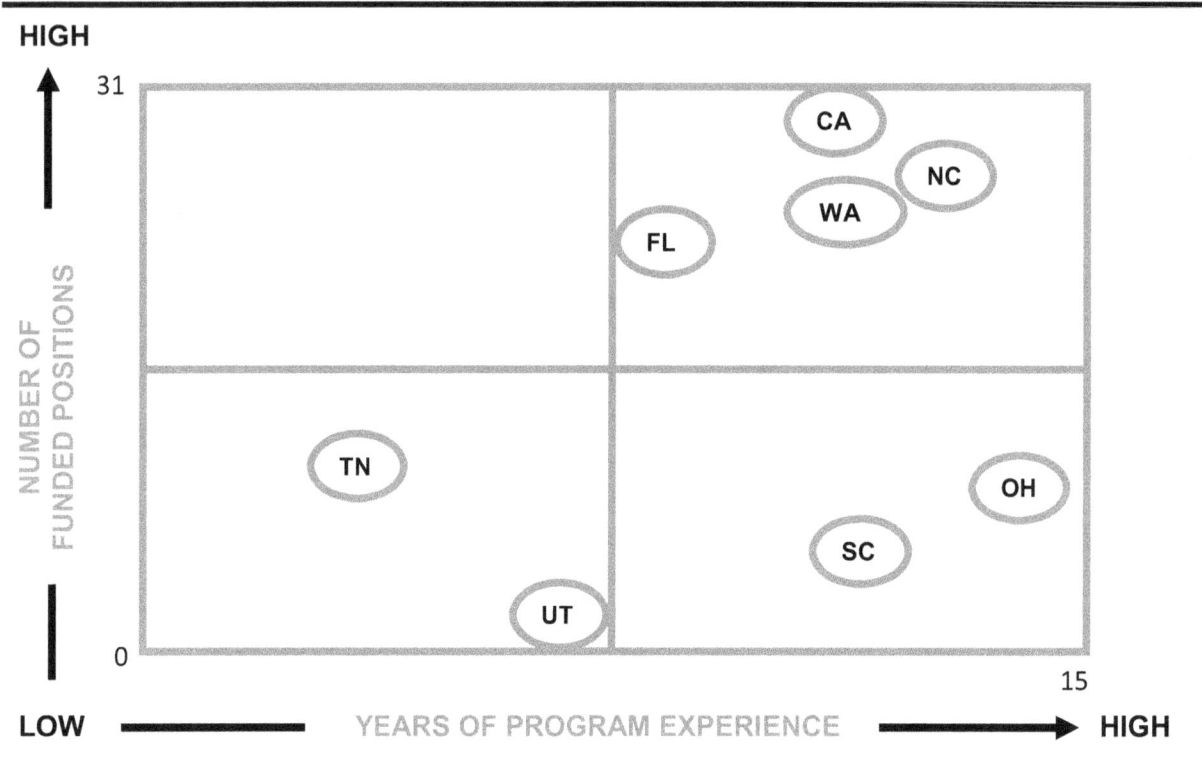

Figure 1. Funded Positions Program Quadrant

Table 3. Funded Positions in Federal Agencies, by State*

State	Federal Agency						Subtotal
	USACE	USFWS	NOAA/NMFS	EPA	USFS	SHPO	
CA	5	5	4	3		3	20
FL	3.5	3	2	2	1	3	14.5
NC		3		2		2	7
OH	4	1		1	1	2	9
WA	3	4	4				11
TN		1		1			2
SC	3	1				1	5
UT							0

*USACE = U.S. Army Corps of Engineers, USFWS = U.S. Fish and Wildlife Service, NOAA/NMFS = National Oceanic and Atmospheric Administration/National Marine Fisheries Services, EPA = Environmental Protection Agency, USFS = U.S. Forest Service, SPHO = State Historic Preservation Office.

Table 4. Funded Positions in State Agencies, by State

State	Water and Coastal Agencies*				Natural Resources, Wildlife, and Conservation Agencies*								State DOT	Other	Sub Total	**Total**
	CA Coastal Commission	FL Water Management Districts	NC DENR Div. of Water Quality	NC DENR Div. of Coastal Management	NC DENR Secretary's Office	NC DENR Div. of Parks and Recreation	NC Wildlife Resources Commission	CA Dept. of Fish and Game	FL Fish and Wildlife Conservation Commission	WA Dept. of Fish and Wildlife	WA Dept. of Ecology	TN Dept. of Environment and Conservation				
CA	5							6							11	31
FL		4.5							2					1[†]	7.5	22
NC			11	4	1	1	2							1[‡]	20	27
OH																9
WA											9				9	20
TN												4	4[§]		8	10
SC														1[#]	1	6

* DNR = Department of Natural Resources, DOT = Department of Transportation. Values represent the number of positions that agencies have been approved to fill; in some cases, these positions are currently vacant.
[†] Position at the Florida Department of Community Affairs.
[‡] Position at the North Carolina Department of Cultural Resources, Office of State Archaeology.
[§] Positions funded by the State Department of Environment and Conservation and located at the State DOT.
[#] Position at the South Carolina Department of Health and Environmental Control.

2.1.1 Program Size

Certain factors, such as the development of performance measures and flexibility with work-task prioritization, were related to a funded positions program's size. Some issues and opportunities unique to larger programs included:

- Difficulty in maintaining consistent communication among liaisons and resource agency/State DOT management, especially in programs where liaisons were located in several different offices or in those where face-to-face meetings were rare.

- Bureaucratic challenges related to obtaining funding for numerous liaisons, as well as the increased administrative burden of processing invoices or financial reports for many funded positions.

- Coordinating among multiple prioritization systems and performance measurements in States where several resource agencies receiving funded positions had independently developed their own systems and measurements.

- Increased opportunities for staff work-task flexibility. One resource agency (with funding for three full-time liaisons) provided funded positions with the flexibility to work part-time on its projects and part-time on State DOT projects. The arrangement was developed as a way to broaden liaisons' work experience. A larger program might have sufficient staff resources to provide liaisons with this option.

Some issues and opportunities unique to smaller programs included:

- Fewer opportunities for staff work-task flexibility. One State DOT and resource agency developed an "exchange" program whereby the State DOT sent a staffer to the resource agency to perform non-State DOT work and the resource agency was tasked with sending a staffer to the State DOT to work on State DOT projects. However, the resource agency lost a staffperson and was unable to send a peer to work at the State DOT. Because there were only a few State DOT staff who dealt with permits, it was difficult for the State DOT to continue to support its staffer at the resource agency without gaining an equivalent peer.

- Increased opportunities to communicate on a one-to-one basis with State DOT or resource agency management. One funded position reported being "overworked and overwhelmed" as the only individual dedicated to transportation projects at a resource agency. However, this liaison also mentioned that he was able to create a close working relationship with agency management; as the sole funded position, he had opportunities to be in constant communication with all resource agency offices.

2.1.2 Program Maturity

There were several differences between funded positions programs that had been in operation for many years and those that had been developed more recently. In general, States with programs that had been in existence for only a few years had fewer funded positions than did those with programs that had been in place for many years. It is possible that difficulties in securing funding and support for the program contributed to these size differences. Many State DOTs and resource agencies reported challenges in justifying the need for program funding to the State legislature and internal administrative State DOT departments. It is also possible that more mature programs with established, demonstrable benefits had less difficulty in obtaining program funding than did more recently established programs. Some States requesting funding to start a program provided very conservative estimates of the number of funded positions required to the State legislature or State DOT administrative departments. These estimates reflected the State's anticipated learning curve with the program and the assumption that funding might need to be increased at a later date when or if additional positions were required. In addition, the estimates often reflected a perceived concern that the State legislature or State DOT administrative departments would not provide a relatively large amount of funding for a program that had not yet led to demonstrable benefits.

Another difference between older and younger funded liaison programs was seen in their development of performance measures. States just starting to implement a program generally focused first on creating quantitative performance measures, since these measures guided liaisons in addressing fundamental program objectives, such as meeting specific project or permitting timelines. In many cases, development of quantitative measures was seen as easier than development of qualitative measures, since quantitative measures were often derived from terms in the funding agreement while typically there was no equivalent existing framework for deriving qualitative measures. Interview findings indicated that performance measures evolved over time; indeed, staff from younger programs reported that the existence of qualitative performance measures usually marked a more mature, experienced program.

Staff from some newer programs reported that they had not yet developed any formal performance measures but that they intended to do so in the future. Some newer programs did not specify any measures in the interagency funding agreements. In other young programs, performance measures were included in the funding agreement but formal assessment of liaisons in relation to those measures had not yet occurred.

States with more mature programs might have benefited from lessons learned over time. Furthermore, these States often perceived a flexibility to experiment with different arrangements for liaison management, program structure, and organization. For example, one resource agency with several years of program experience had experimented with several organizational setups to determine who would manage the funded positions and how many positions the manager would oversee. The agency's original eight funded positions had first been physically located in different offices with different supervisors. Later, the liaisons relocated to the same office. The agency discovered that keeping the positions in the same physical space with the same supervisor was the most efficient and effective arrangement. This structure also facilitated communication between liaisons and their supervisor. More recently established programs might not yet have had opportunities to benefit from lessons learned or to experiment with different organizational arrangements.

Finally, staff in more mature programs often reported on both short- and long-term benefits of having funded positions, highlighting the liaisons' ability to facilitate communication between the resource agency and the State DOT while improving relationships among agencies. On the other hand, staff at some younger programs mentioned that it was too early to tell the exact benefits that funded positions had provided. As good working relationships between resource agencies and State DOTs took time to develop, it is likely that some younger programs may not have been in existence long enough for staff to derive long-term positive benefits.

2.2 Overview: The Decisionmaking Process

State DOTs and resource agencies make a number of important decisions during each stage of developing, implementing, and managing funded positions programs. Broadly, the end-to-end process, from creating a funded positions program to assessing its benefits, can be divided into six discrete stages:

- **Stage 1:** Assessing the need and demand for funded positions
- **Stage 2:** Generating program support
- **Stage 3:** Designing a funded positions program
- **Stage 4:** Formalizing interagency agreements
- **Stage 5:** Implementing and managing the program
- **Stage 6:** Evaluating program outcomes

Figure 2 provides a graphic framework of the stepwise process for considering individual decisions within each of the six decisionmaking stages. The left portion of the figure depicts important decisions considered within each stage, and the right side shows additional factors for deliberation. Funded positions programs at various State DOTs and resource agencies may fall under an earlier or later stage of development; the framework allows each agency to identify where it is in the process, and to use the questions to tailor an approach to assessing its own program.

Stage 1: Assessing the need and demand for funded positions

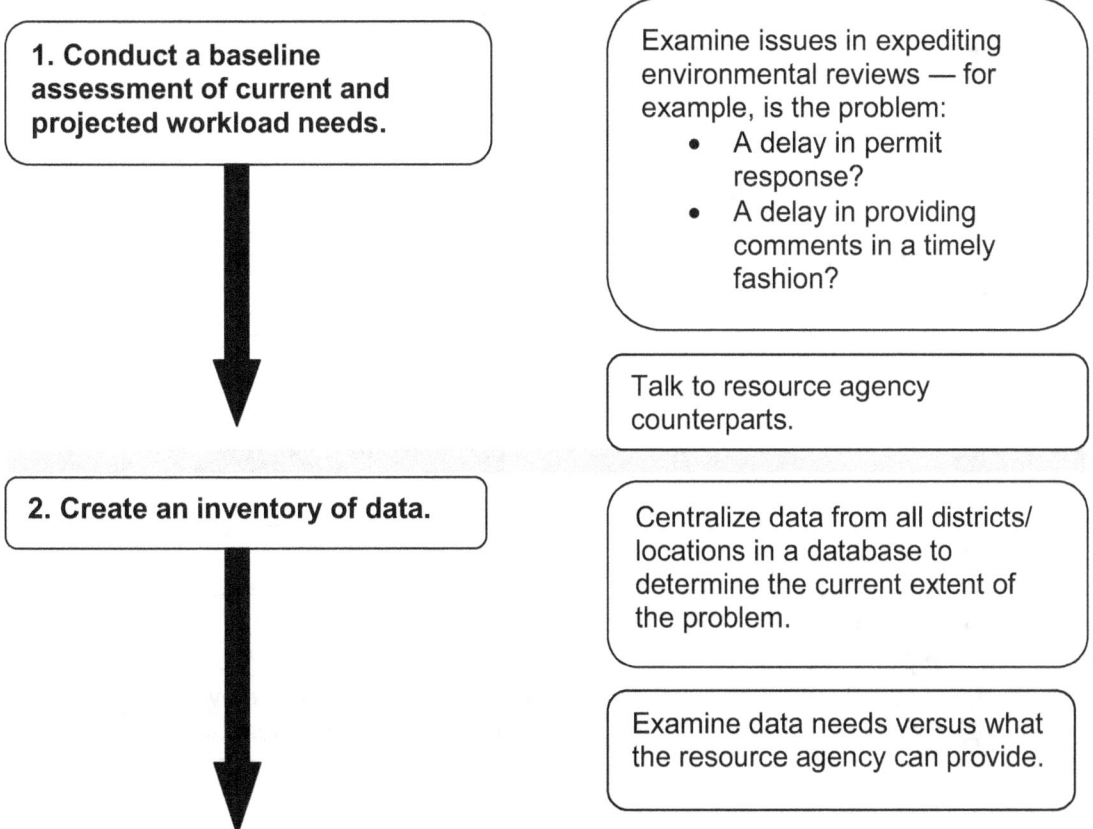

Figure 2. Six-Stage Decisionmaking Process

Stage 2: Generating program support

1. Gain support from upper-level management/State government officials.
   - Present results of successful existing programs.
   - Focus on the benefits of facilitating reduced permit-turnaround times.

2. Implement a "pilot" funded positions program.
   - Test the process via a streamlining agreement.
   - Start by requesting a small number of positions.

Stage 3: Designing a funded positions program

1. Consider where to locate the funded positions.
   - Allow funded positions to work some of the time at the State DOT to help build relationships.

2. Ensure that term length is long enough to attract and keep qualified candidates.
   - Examine agency FTE ceilings and determine whether the position can be permanent.
   - Make term lengths long enough to avoid excessive turnover (five years is optimal).

3. Determine appropriate grade level and level of experience.
   - Give liaisons more ability to mediate/negotiate by hiring individuals at higher grade levels and those with more experience.

Figure 2. Six-Stage Decisionmaking Process (continued)

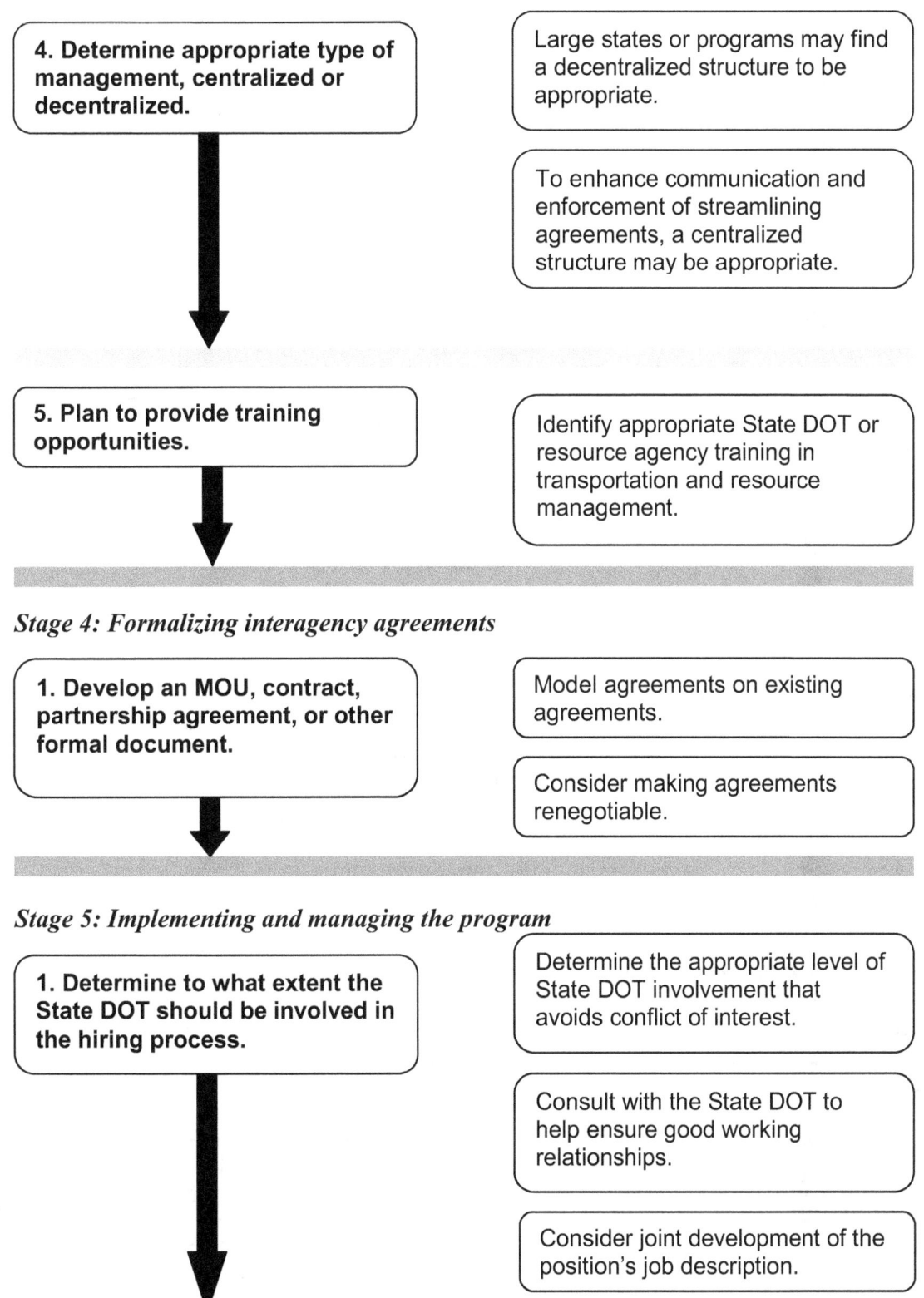

Figure 2. Six-Stage Decisionmaking Process (continued)

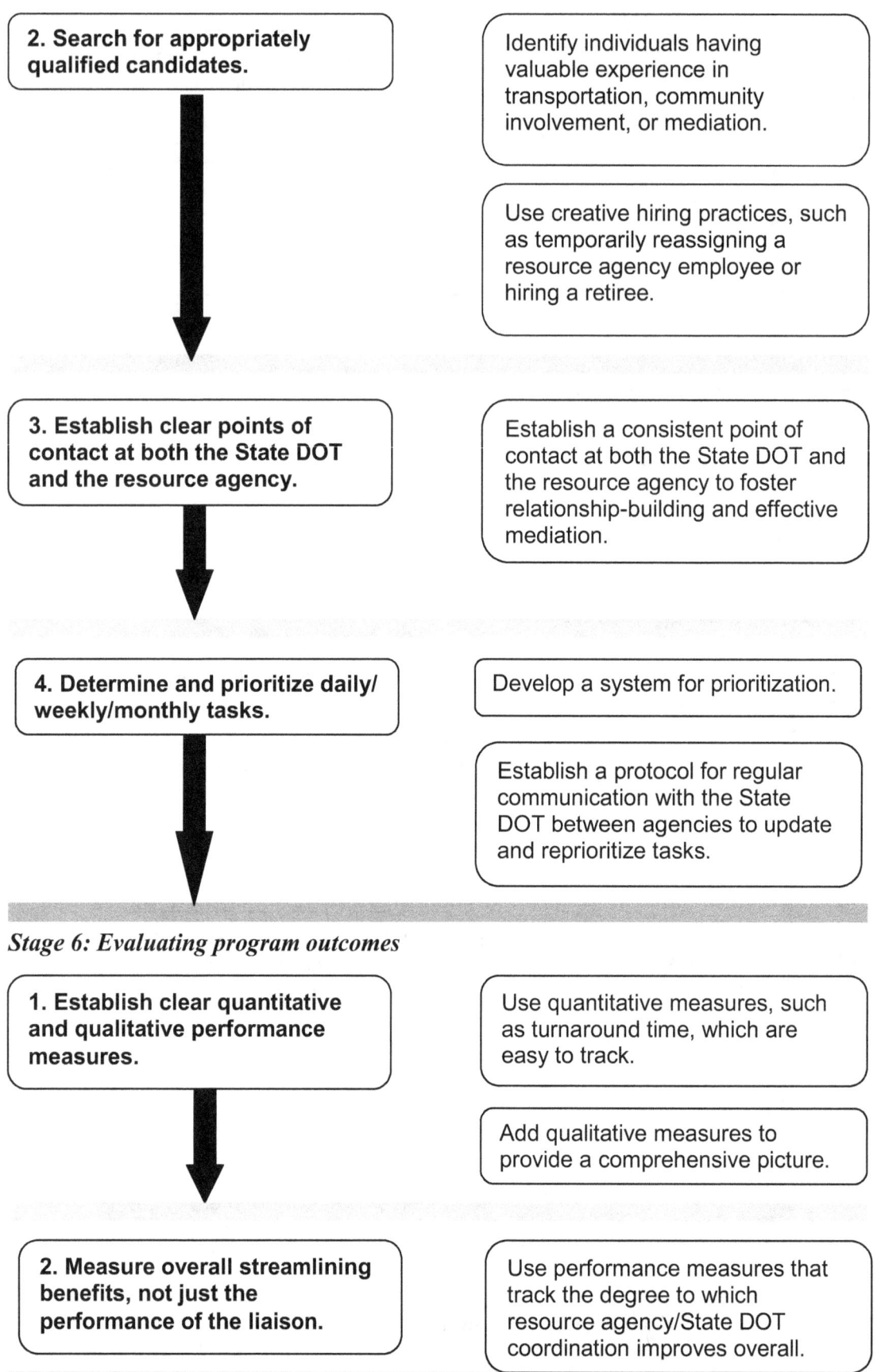

Figure 2. Six-Stage Decisionmaking Process (continued)

3 Stage 1: Assessing the Need and Demand for Funded Positions

This section explores the common areas of demand identified when initiating funded positions programs and the ways in which agencies have assessed the need for these programs. It is clear that demand for funded positions has increased as their value has become apparent: in 2003, there were 222 positions in 18 States, while in 2005, there were 375 funded positions in 34 States. One-third of the funded positions were located at Federal agencies, and two-thirds were at State resource agencies.[8]

As demand for funded positions has increased, some agencies have justified adding new positions by conducting baseline assessments of current and projected workload needs.

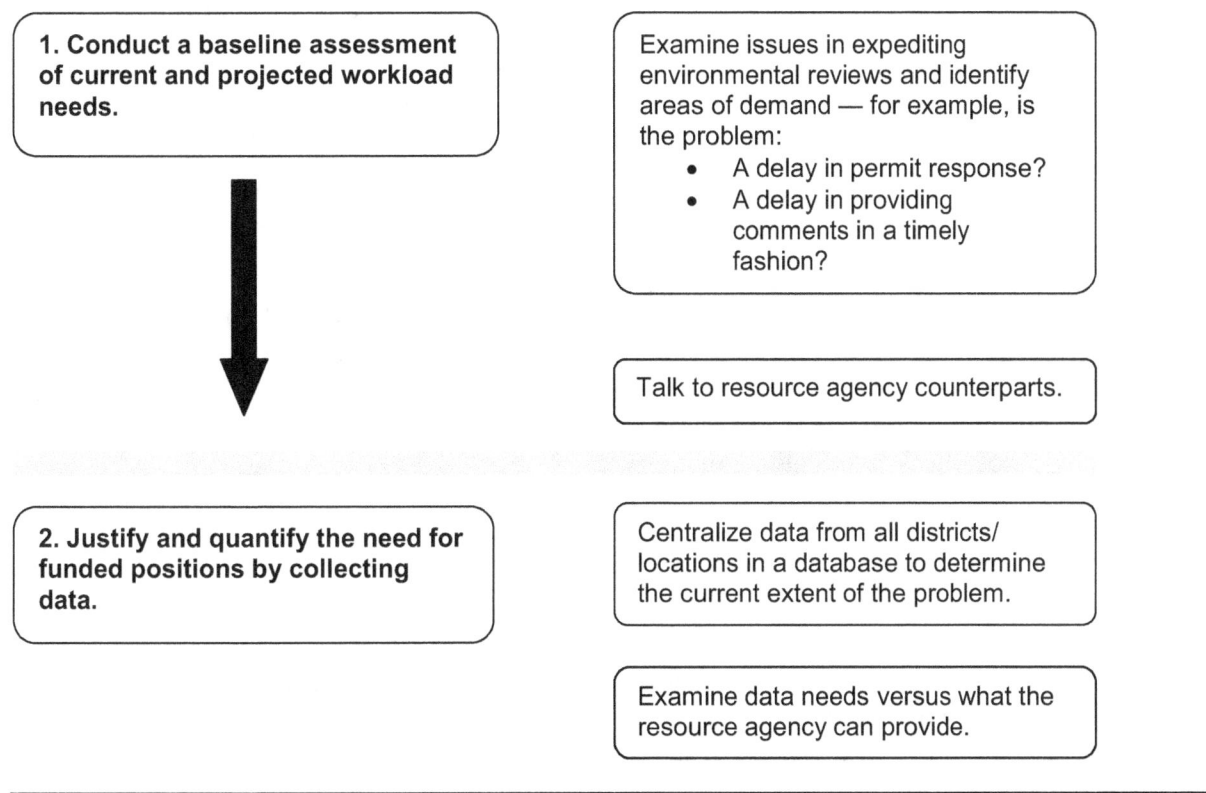

Figure 3: Assessing the Need and Demand for Funded Positions

3.1 Conducting a Baseline Assessment

3.1.1 Identifying Areas of Demand

Various factors have influenced the increasing demand for funded positions. Some of the issues identified by State DOTs and resource agencies that led to the development of funded positions programs were:

---

[8] AASHTO Center for Environmental Excellence, May 2005.

- **Increased work demands.** Resource agency staff had to address both an increase in the volume of transportation-related work and the constraint of having insufficient staff to meet new work demands.

- **Need to expedite project delivery.** AASHTO's May 2005 report found that, of the 275 positions, all but 15 were devoted to project delivery.[9] Interviews also showed that project delivery was an important aspect of funded positions' responsibilities. Several states reported that demand for funded positions was directly related to difficulty in meeting stated timelines for project review, resource agency difficulty in providing quality deliverables to State DOTs, or an increased number of transportation projects in the pipeline and/or pressure to reduce project timelines.

- **Need for premium level of service.** Some State DOTs and resource agencies preferred to have dedicated staffpersons at the resource agency to focus on transportation-related projects.

- **Need to increase staff resources.** Additional staff resources were often required to address new regulatory processes, such as SAFETEA-LU's emphasis on planning.

- **Desire to increase access to experts.** Funded liaisons were sometimes viewed as expert staff who could augment resource agency knowledge or provide guidance on certain activities and processes.

- **Need to improve working relationships.** Some funded positions were utilized to establish more effective dialogue between resource agencies and State DOTs or to facilitate earlier resource agency involvement with State DOTs.

- **Need for support on a wide range of environmental regulations.** Funded positions provided regulatory support, such as reviewing permitting, review, and consultation requirements under laws including Endangered Species Act Section 7, Clean Water Act Sections 404 and 401, National Historic Preservation Act Section 106, U.S. Coast Guard bridge clearances, and State regulations.

- **Need for "big-picture" support.** Some State DOTs and resource agencies used funded positions to promote a programmatic rather than a project-by-project approach.

- **Desire for other services.** Some agreements provide funding for services rather than positions. MaineDOT's agreement with the Maine Historic Preservation Commission covers funding for services performed, such as archaeological services and the cataloguing and mapping of archaeological sites. California's agreement with USFWS also specifies services including "environmental technical assistance, consultation, and coordination services related to the Department's projects, on an on-call basis."

For many State DOTs and resource agencies, problems concerning project delivery and permitting timeframes helped to demonstrate the need for increased staffing or staff dedicated to State DOT projects at the resource agencies. However, most State DOTs or resource agencies did not ascertain whether other mechanisms besides funded positions programs, such as developing service contracts or making process improvements, would help to meet these needs.

---

[9] AASHTO, *op. cit.*, p. 6.

3.1.2 Communicating with the Resource Agency or State DOT

Interviewees reported several models for identifying the need for a funded positions program and communicating it to the resource agency or State DOT. The primary models were:

- **Model A:** The State DOT identified a difficulty with moving projects forward and required focused resource agency involvement to assist in decisionmaking.

- **Model B:** The resource agency identified many competing priorities and a lack of staff resources and reported these needs to the State DOT.

- **Model C:** The State DOT and several resource agencies simultaneously came to similar conclusions about the need for liaisons and conducted meetings to discuss implementation of a multiagency liaison program.

Each of the models is illustrated and explored in more detail below. Critical factors for all three models included a willingness to engage as well as the establishment of open lines of communication at management levels of both agencies.

A.

In situations where the State DOT initiated implementation of a funded positions program, it reported that a primary motivator was the need to expedite the resource agency's review of State DOT project permits. Another primary driving factor was the State DOT's need for increased access to resource agency experts. State DOTs viewed the funded positions program as a mechanism to increase dedicated staff support at resource agencies and improve the transportation decisionmaking process.

B.

In some cases, one resource agency realized that the demand for staff time exceeded staff availability or that an increase in staffers was required to provide an improved level of service to the State DOT. The resource agency communicated these needs to the State DOT and suggested that funded liaisons could help to augment staff time and availability.

C.

Other scenarios involved discussion about funded positions among multiple resource agencies, leading to conversations with the State DOT. Similarly, a series of dialogues between the State DOT and several resource agencies could occur regarding a need to make process improvements or to develop an environmental streamlining agreement. In some of these cases, the initial focus of the conversations might have been on the resource agency's needs and negotiating the number of liaisons to fund. However, the process of engaging in dialogue also led to outcomes such as the development of regular meetings between the State DOT and resource agencies. These types of outcomes benefited all stakeholders by improving State DOT-resource agency communication and allowing all parties to feel a sense of ownership of the process.

When assessing the need for a funded positions program, the resource agency and the State DOT should have clarity on what is required from the liaison. It is also important to come to agreement on these needs before establishing or implementing a program. Agreeing on agency needs as well as program goals and objectives early on when developing a program can help to establish a framework for continued trust, collaboration, and effective communication throughout the process.

3.2 Justifying and Quantifying the Need

For many State DOTs and resource agencies, problems concerning project delivery and permitting timeframes helped to demonstrate the need for increased staffing or staff dedicated to State DOT projects at the resource agencies. However, most State DOTs or resource agencies did not ascertain whether other mechanisms besides funded positions programs, such as developing service contracts or making process improvements, would help to meet these needs. In one instance, a State DOT reported that a resource agency had approached it to request funding for a liaison. The State DOT examined the projected labor hours for the liaison and determined that there was not a true need to add a staffperson at the resource agency. The State DOT decided not to fund a position, instead implementing an "exchange" program to locate a State DOT employee part-time at the resource agency to work on non-State DOT projects; in exchange, the resource agency would provide a peer to work at the State DOT on State DOT projects.

Several State DOTs and resource agencies successfully justified funding liaisons by documenting workload increases or providing data on the numbers of funded positions needed. Here are some examples:

- One resource agency presented the State DOT with a spreadsheet listing the number of State DOT projects that resource agency staff were currently working on relative to the number of projects being worked on for other customers.

- Another resource agency determined the number of positions needed by deciding how many liaisons it could employ on a permanent basis, using appropriations monies if funding became unavailable. The State DOT then funded only as many positions as the resource agency determined it could permanently support.

- Several State DOTs calculated the workload and the number of liaisons requested by assessing the number of permits needed as well as performance data for each of the resource agencies.

- One State DOT provided initial documentation to the resource agency that substantiated the need for liaisons. The resource agency then reviewed and added details to the documentation.

During the renegotiation of existing funding agreements, a State DOT specifically asked each resource agency to justify its request for a specific number of funded positions. Justifying information included the amount of State DOT work at the resource agency and the type of support the resource agency required to meet these work demands.

Recommendations: Assessing the Need and Demand for Funded Positions

Before establishing a funded position, a State DOT and a resource agency should have clarity on the need for a funded positions program and should agree on this need. In addition, the State DOT and the resource agency should consider different methods for communicating the need to each other. Discussions about need could be initiated by the resource agency, the State DOT, or collaboratively by both agencies.

To determine whether a need for funded positions exists, the State DOT and the resource agency should consider the following:

- Are there increasing work demands and an inability to meet them, including an inability to provide deliverables within stated timeframes?
- Are there new work processes that require closer collaboration between the State DOT and the resource agency and that are likely to slow existing projects?
- Is there contention between the State DOT and the resource agency, or are there community difficulties?
- Would the addition of a dedicated staff member help to expedite resource agency reviews of State DOT projects or to better address State DOT concerns?

To justify and quantify the need for funded positions, agencies should collect data such as:

- Baseline measurements of current workload demands.
- Number of permits required within a specified timeframe.
- Past performance data on State DOT permitting.
- Number of liaisons that a resource agency could absorb on a permanent basis if a term position became unavailable.
- Number of labor-hours spent working on transportation projects as compared with other resource agency projects.

4 Stage 2: Generating Program Support

Generating support for a funded positions program requires more effort than simply quantifying the need for funded positions; it also requires a demonstration that the program will produce positive and substantial improvements in project review and delivery.

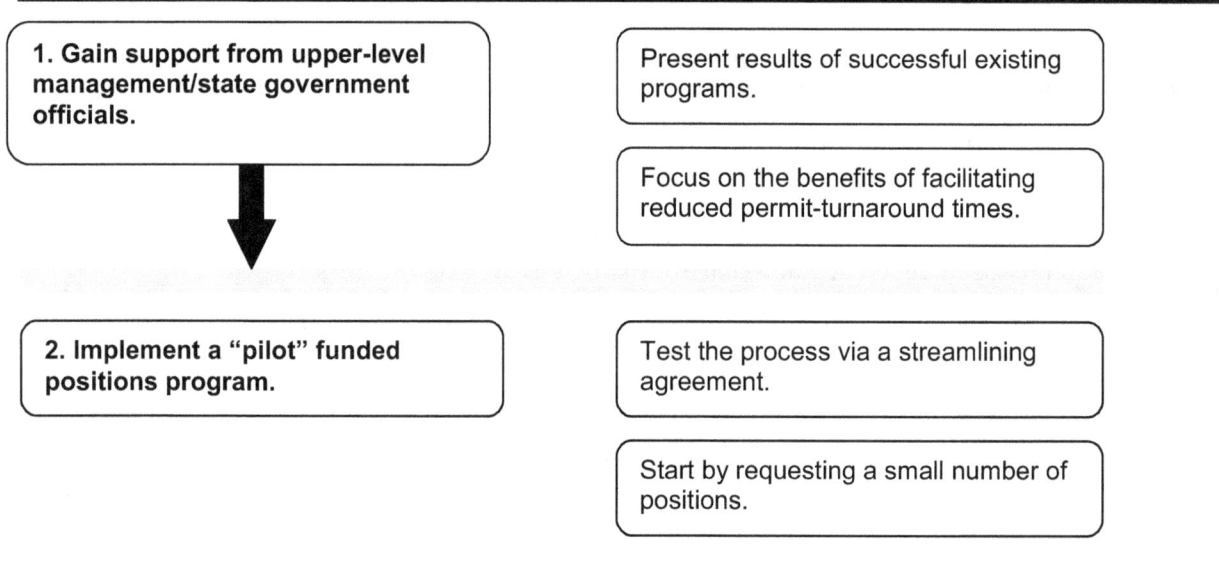

Figure 4. Generating Program Support

4.1 Gaining Support from Management and Government Officials

The results of successful existing funded positions programs can be extremely helpful when seeking support for establishing or expanding a funded positions program. In AASHTO's May 2005 report, many State DOTs reported benefits from the use of funded positions at resource agencies.[10] An NCHRP Peer Exchange report on funded positions summarizes several benefits identified by 15 individuals representing FHWA headquarters and field offices, State DOTs, and resource agencies.[11] This review of the general literature found the following:

- **State DOTs and other agencies are satisfied with the support they receive from funded positions.**[12] They often state that the positions help the environmental review process to work more smoothly.

- **Funded positions increase predictability and decrease turnaround times of projects and reviews.** Ohio DOT quantified the benefits of funded positions and found that the number of planned projects delivered within two weeks of their scheduled date increased from 70 percent in 1999 to 96 percent by 2007. Ohio DOT also reported cost savings of

---

[10] AASHTO, *DOT-Funded Positions and Other Support to Resource and Regulatory Agencies, Tribes, and Non-Governmental Organizations for Environmental Stewardship and Streamlining Initiatives,*" p. 30, May 2005.
[10] *Meeting 2: Work Assignments and Performance in External Positions Funded by State DOTs.* NCHRP Project 08-36, Peer Exchange Series on State and Metropolitan Transportation Planning Issues, pp. 3-5–3-6, December 2007 (unpublished),.
[11] NCHRP Peer Exchange Report, *op. cit.*, pp. 2-7–2-8.
[12] AASHTO, *op. cit.*, p. 30.

$100,000 and time savings of four to six months on the NEPA process for large projects and cost savings of $30,000 and two to three months for small projects.[13]

- **Funded positions increase capacity and value.** With many resource and regulatory agencies facing budget constraints, the liaisons have been essential for ensuring responsive support for environmental review activities.

- **Funded positions foster positive interagency relationships.** The liaisons created a bridge between organizations whose missions and values often differ.

4.2 Implementing a Pilot Program

Another way to garner support for a new funded positions program is to propose a temporary "pilot" program that tests the program objectives with just a few funded positions. The pilot program should have clear performance objectives (see Section 8 of this report for more on this subject), which, if met, would justify increased investment in funded positions in the State.

Recommendations: Generating Program Support

Gaining support from higher-level officials: Presenting results of successful existing funded positions programs can be extremely helpful when seeking support for establishing or expanding a funded positions program. Focusing on decreased permit-turnaround times helps to make the benefits of the program concrete.

Implementing a pilot program: A good starting point when requesting funding is to propose a temporary "pilot" program that tests the program objectives with just a few funded positions. Pilot programs should have clear performance objectives; this will allow the resource agency and the State DOT to clearly demonstrate funded positions' benefits.

---

[13] NCHRP Peer Exchange Report, *op. cit.*, pp. 3-5–3-6.

5 Stage 3: Designing a Funded Positions Program

This section deals with the logistics of a funded positions program, including elements such as the location of funded positions, determination of term lengths, requirements for levels of experience, centralization of management, and training. State DOTs and resource agencies should consider each of these elements within their individual contexts before proceeding with an interagency agreement.

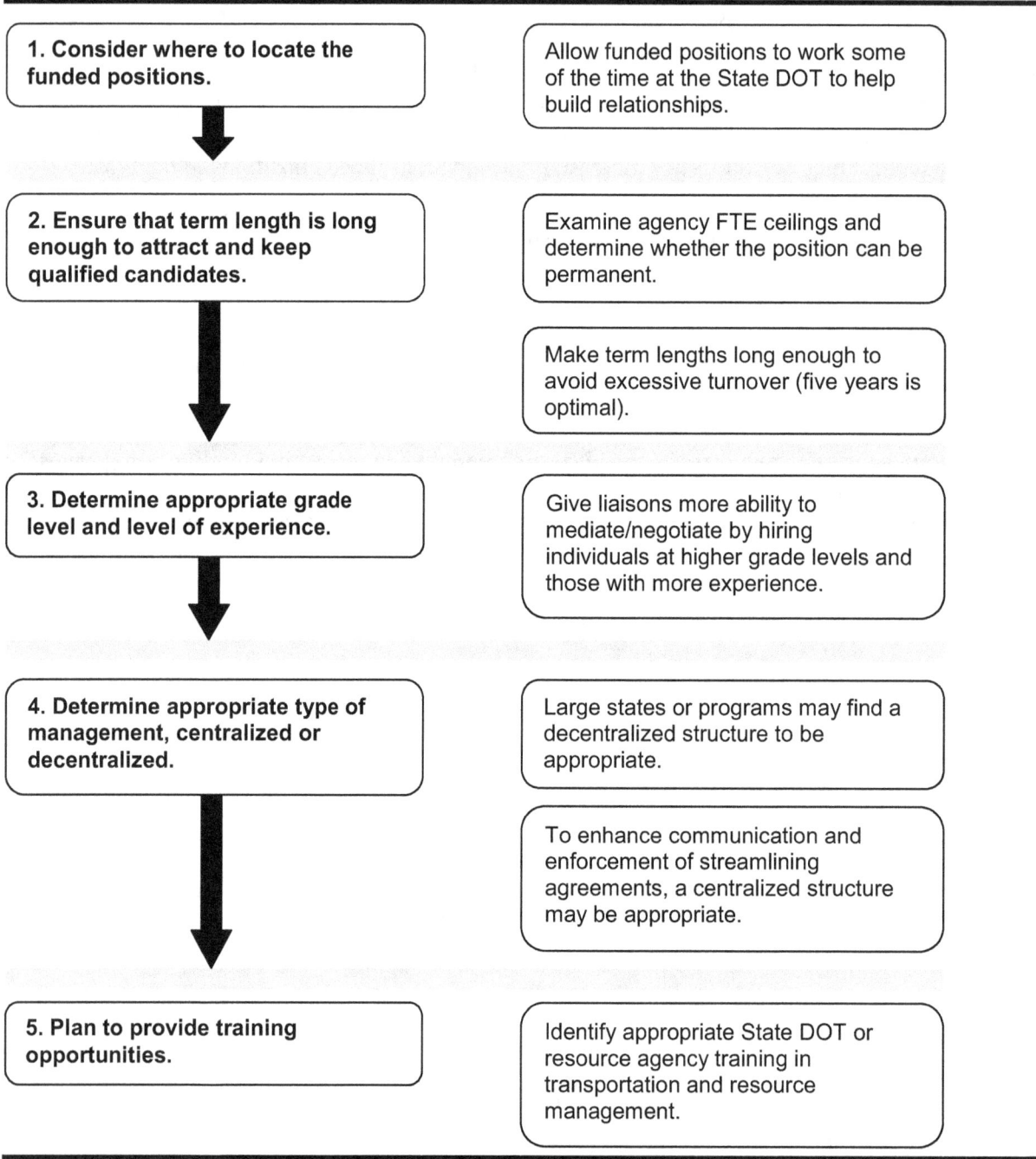

Figure 5. Designing a Funded Positions Program

5.1 Locating the Funded Positions

The physical location of the funded positions appeared to have some effect on the quality of communication between the funded positions and the State DOT. For some States, physically locating a funded position at the State DOT, even on a part-time basis, facilitated stronger working relationships. In one instance, a State DOT prepared an office for the liaison to work in occasionally; the liaison reported that this led to positive outcomes. As a result of working at the State DOT, for example, the liaison was able to learn more about its projects and to exchange information with it on a timelier basis. While this arrangement was beneficial in this particular circumstance, another resource agency expressed concern that, if physically located at the State DOT, the liaison might not be able to obtain sufficient resource agency support.

"We went back and forth on where to locate the funded position: near the client [DOT] or near the mothership [resource agency]. We ended up locating the position at the resource agency. We discussed [this arrangement] with the [State] DOT and they didn't express any strong opinions. As long as the person actively participates in the process the [State] DOT was okay with wherever they sat."
— Resource agency

In several instances, resource agencies were reluctant to physically locate the funded position full time at the State DOT offices. It is likely that these agencies wanted to ensure that the liaisons were clearly identified as employees of the resource agency rather than of the State DOT.

5.2 Determining Term Lengths

The NCHRP Peer Exchange report found that "a challenging component to managing external positions is their short duration, typically one year, which makes it difficult to market the position as an attractive option to potential qualified candidates. Moreover, the short duration of externally funded positions results in a large amount of paperwork."[14] The duration of the term is therefore an important consideration when establishing a funded position, in terms of both paperwork and attracting qualified candidates. Highly qualified candidates are less likely to apply for a position that does not promise ongoing employment.

Table 5 displays the varying term lengths for funded positions in 46 agreements.

Table 5. Duration of Service for Funded Positions

Duration (yrs.)	No. of Agreements	Percentage	States
5	10	22%	FL, TN, IN, AR
4	1	2%	MT
3	12	26%	FL, KY, LA, MS, MT
2	9	20%	FL, ME, TX
1	6	13%	ME, PA
Unspecified	8	17%	AR, AZ, CA, ID, NJ, WA
Total	46	100%	

---

[14] NCHRP Peer Exchange Report, *op. cit.*, p. 3-2.

5.3 Determining Experience and Grade Levels

The agreements in the literature review specified funded positions with various grade levels ranging from GS-9/11 to GS-13. Tennessee's agreement asked for Environmental Specialists at the "3, 4, and 5" levels. Other agreements specified senior-level staff; for example, Florida's agreement with the Jacksonville District, USACE, stated that the "Corps is assigning senior, experienced employees to perform these tasks …". Only the agreement of the Louisiana Department of Transportation and Development (LA DOTD)–Louisiana Department of Culture, Recreation and Tourism, Office of Cultural Development, Division of Archaeology, stipulated a position at the "Student Worker" level.

Although it may be less expensive to hire a funded position with a lower grade level, study participants found that experienced or senior-level employees often provided more abstract benefits beyond streamlining the permitting process, such as helping to initiate and improve processes or providing input on higher-level policy issues.

5.4 Examining Centralized and Decentralized Management

The advantages of centralized versus decentralized management will be discussed more fully in Section 8 of this report. Before creating an interagency agreement, however, it is worth considering whether to allocate funds for a central coordinator for all funded positions in a particular State. The advantages of central coordination can include greater efficiency and prevention of duplicated efforts. However, for some programs it may be more appropriate to disperse management and have each funded position report to a district-level office.

5.5 Providing Training Opportunities

Many liaisons had expertise in their particular resource fields but lacked extensive knowledge of transportation issues and processes. Most State DOTs included training funds as part of the funding agreement, and several offered courses to new-liaison hires that specifically focused on the transportation decisionmaking process.

Of the interagency agreements in the literature review, 20 (in the States of Arkansas, California, Florida, Montana, and Tennessee) stipulated that training for the funded position would be provided. However, in 22 of the agreements, funded positions were required to coordinate or provide training in their areas of expertise for the State DOTs. (There was an overlap of 10 agreements from Florida that specified both training for the funded positions as well as requirements for the funded positions to train State DOTs.)

> *"The learning curve was enormous — you're thrown in a room with fifty people, and maybe five or six of them are from resource agencies. You are talking traffic counts, etc. — it's a whole new language."*
> – Funded position

State DOTs viewed liaisons as valuable sources of knowledge on resource agency practices. Several funded positions were tasked with providing training to State DOT staff on improving compliance with resource agency requirements and processes.

In some cases, the State DOT provided funding for the liaison to attend non-State DOT-hosted trainings. In addition to transportation-related training, most liaisons received instruction on the resource agency's regulatory responsibilities. AASHTO's 2005 survey found that many State DOTs offered funding for training, conferences, and other professional-capacity activities. State DOTs relied heavily on classes provided by the National Highway Institute.[15]

Some common training topics provided by the State DOT included:

- Context Sensitive Solutions (CSS),[16] an interdisciplinary approach that involves all stakeholders in providing a transportation facility that fits its setting.
- NEPA and cross-cutting environmental issues.
- Environmental review improvement through collaborative problem-solving.

Examples of specific training approaches were as follows:

- A State DOT included a stipulation in the funded positions' contracts that liaisons would dedicate 10 percent of their time to training activities.
- A State DOT and a resource agency worked together to develop an orientation course to help introduce new funded positions to their roles and responsibilities as well as to State DOT and resource agency processes.
- One resource agency implemented a mentor system to orient new liaisons to the program. This system helped the liaisons to gain a better understanding of the resource agency's culture and mission.
- One State developed informal opportunities for liaisons to network and interact with each other. Such opportunities provided low-cost ways to increase information-sharing and provide on-the-job learning.
- In one case, a funded position reported that the State DOT would not directly fund the individual to attend training but would cover the labor-hours required for attending training sessions.

---

[15] AASHTO, *op. cit.*, p. 24.
[16] For more information, see www.contextsensitivesolutions.org/content/topics/what_is_css/.

Recommendations: Designing a Funded Positions Program

Locating funded positions: Physically locating a funded position at the State DOT, even on a part-time basis, can facilitate stronger working relationships.

Determining term length: The longer the term of service that a funded position is guaranteed, the more likely that the position will attract highly qualified candidates. Shorter term lengths make the hiring process more difficult.

Determining grade level or level of experience: Funded positions at lower grade levels will require less funding. However, experienced or senior-level employees often provide more abstract benefits beyond streamlining the permitting process, such as helping to initiate and improve processes or providing input into higher-level policy issues.

Centralizing versus decentralizing management: Agencies should also evaluate the program to determine whether a centralized or decentralized management structure would produce better results. Factors involved in this decision include:

- The size of the program and the geographic distance between offices. Large programs or large States may make centralized management impractical. However, even decentralized programs should make provisions for liaison-to-liaison communication.

- The potential for regional coordinators or central offices to streamline implementation of the funded positions agreement. Having a centralized management structure can allow some staff to focus on monitoring the terms of the agreement while others focus on engaging in tasks related to the funded positions program.

Providing training: State DOTs and resource agencies should determine what training opportunities they could provide to meet funded positions' needs. Different training approaches can be adopted to meet the liaison's skill set and the specific circumstances of the State DOT and the resource agency. These approaches can be developed by the State DOT, the resource agency, the funded liaison, or collaboratively by all parties. Appropriate methods of training include:

- Developing orientation training for all new funded positions, possibly to include a "mentor" system for existing liaisons.

- Encouraging networking opportunities for liaisons to interact with one another.

- Funding liaisons to attend relevant external conferences or courses at training centers.

Training topics: Topics for training could include Context Sensitive Solutions (CSS), NEPA, or improving environmental reviews. In addition to providing training to funded liaisons, State DOTs and resource agencies can consider whether the liaison should provide training to the agencies.

6 Stage 4: Formalizing Interagency Agreements

This section details how agencies document position needs in interagency agreements, such as Memoranda of Understanding (MOUs), contracts, partnership agreements, or other formal documents. It also discusses the ways that State DOTs and resource agencies have been able to renegotiate agreements on the basis of changing financial, organizational, and legislative environments.

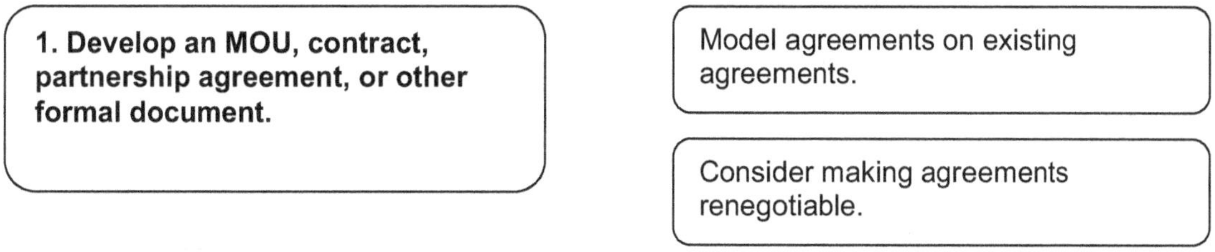

1. Develop an MOU, contract, partnership agreement, or other formal document.
- Model agreements on existing agreements.
- Consider making agreements renegotiable.

Figure 6. Formalizing Interagency Agreements

6.1 Modeling Agreements on Existing Agreements

Various agreement mechanisms are used to formalize funding agreements between State DOTs and agencies. Because State law regarding interagency agreements varies, State DOTs and resource agencies should examine existing interagency agreements from their State, paying special attention to the funding mechanisms. An examination of funded positions agreements from other States is also valuable for identifying best practices.

6.1.1 Characteristics of Existing Agreements

To demonstrate the high level of variability among existing agreements, a few examples of innovative and unusual interagency agreements are highlighted below:

- AASHTO reported that "a small number of states used Intergovernmental Personnel Act (IPA) agreements, particularly for staff at federal agencies."[17] These agreements often have term limits, while others have enabled staff to be kept on for ten years or more.

- Missouri, Minnesota, and Texas are prevented by State law from funding positions at other agencies. In Texas, FHWA and TxDOT agreed to use some discretionary funds on behalf of TxDOT in order to fund positions at resource agencies.

- Florida DOT (FDOT) has developed master agreements to fund positions. These agreements define how the State will implement its Efficient Transportation Decision-Making (ETDM) process.

- FDOT and North Carolina DOT (NCDOT) have developed funded-positions manuals to guide the development and management of funded positions and related activities.

- Most agreements that mention funded positions' work responsibilities have done so in a separate appendix containing a short checklist. For example, Arizona DOT's (ADOT)

---
[17] AASHTO, op. cit., p. 19.

agreement with USACE provided an appendix that detailed the NEPA–Section 404 individual permit process. The appendix gave specific instructions on the documents and materials that each agency was to provide the liaison and vice versa.

Existing agreements also vary in terms of the number of positions funded, the funding amounts allocated, and the sources of funding. Tables 6 through 8 display the characteristics of 46 agreements examined in 2008 during this study.

Table 6. Number of Positions Funded per Agreement

No. of Positions	No. of Agreements	Percentage (of all Agreements)	States
3–9	8	17%	FL, TN, WS[18]
2	8	17%	FL, IN, LA, PA
1–1.5	15	33%	AR, AZ, FL, KY, LA, MS, MT, PA
Unspecified	15	33%	AZ, CA, FL, ID, ME, NJ, PA, TX
Total	46	100%	

Table 7. Funding Amounts per Agreement

Funding	No. of Agreements	Percentage (of all Agreements)	States
> $1 million	9	20%	CA, FL, TN
$500,000–$1 million	4	9%	FL, WA
< $500,000	25	54%	AR, FL, KY, LA, ME, MS, MT, PA, TX
Unspecified	8	17%	AR, AZ, CA, ID, NJ, WA
Total	46	100%	

Table 8. Funding Sources for Funded Positions

Source	No. of Agreements	Percentage (of all Agreements)	States
100 percent federal funding	7	15%	FL
80/20 (federal/State or other)	13	28%	AR, IN, LA, ME, MS, MT,[19] PA, TN
Unspecified	26	57%	AR, AZ, CA, FL, ID, KY, LA, MT, NJ, PA, TX, WA
Total	46	100%	

---

[18] FDOT–USFWS had an agreement that mentioned three positions, TDOT–Tennessee Department of Environment and Conservation's agreement specified four positions, WSDOT–Washington State Department of Ecology's agreement covered a 4.15 biennial FTE, and FDOT–Jacksonville District USACE's agreement covered nine positions. WSDOT's agreement was unique in that it listed the names of the specific Ecology Project liaisons that it was funding.

[19] Montana's agreement stated that there would be a Federal-aid match rate of 86.58 percent, with the rest coming from non-MDOT State funds.

Most of the larger funding amounts were found in agreements that covered longer durations (three or more years). It is not clear whether the larger budgeted sums also correlated with a greater number of positions per agreement. A third of the agreements that were examined did not specify the number of positions that were funded with the money that was allocated. For example, one agreement had a specific funding amount for three years, but did not indicate the number of positions that were being funded.

6.1.2 Best Practices in Existing Agreements

The analysis of interagency agreements identified characteristics of good practices as follows:

- **Provide guidance about priorities.** For example, Montana DOT's (MDT) agreement with USFWS states that "on a quarterly basis, MDT will establish and submit a priority listing of projects to USFWS which will guide USFWS supplemental staff efforts in the priority review process."

- **Reference specific regulations.** The California Department of Transportation's (Caltrans)–USFWS agreement includes an MOU that addresses specific environmental streamlining regulations.

- **List clear performance objectives.** FDOT's many agreements have clear language on review timelines. The Montana Fish, Wildlife and Parks agreement has an entire section on performance objectives, which are operational and measurable. Similarly, Tennessee DOT's (TDOT) agreement with FHWA and the Tennessee Department of Environmental and Conservation contains sections on both agency coordination and performance objectives," with specific requirements delineated for the project-concurrence process, training classes, and permit-application-review timelines.

- **Insist on cooperation and collaboration**. The Louisiana Department of Transportation and Development's (LA DOTD) agreement with USFWS specifies that "all signatory agencies agree that ready and reasonable access will be provided to working level staff of the other agencies in an effort to minimize the need for formal meetings."

- **Make training a strategic priority.** Montana Department of Transportation's (MDT) agreement with FHWA and the State Department of FWP provides for training for the position to allow familiarization with program requirements.

- **Focus on coordination.** The section on scope in the Kentucky Transportation Cabinet's (KYTC) agreement with USFWS requests that "initially, KYTC and the USFWS will hold monthly meetings to discuss coordination of the priority review process." The MDT–USFWS agreement has similar language. Coordination is one of the most important (yet implicit) purposes of having funded positions.

- **Specify responsibilities for each agency.** Some agreements, including those of Texas DOT (TxDOT), utilize a format that lists all agencies' responsibilities, not only those of funded positions.

6.2 Renegotiating Agreements

Agreements should be viewed as ongoing collaborations between State DOTs and resource agencies rather than as the static responsibility of a single agency. Given the changing needs that State DOTs and resource agencies are likely to face as regulations and projects change, study participants expressed frustration that States with multiple State DOT–resource agency funding agreements tended to renegotiate these agreements on an ad hoc basis. It was difficult for State DOTs to simultaneously renegotiate all statewide agreements. In addition, agreements were sometimes developed at different times and might not have been ready for renegotiation at the same time. The streamlining of agreements by creation of a master agreement that is renegotiable in terms of funding and number of positions funded can ultimately lead to a much less cumbersome process.

Recommendations: Formalizing Funding Agreements

Formalizing the funded positions program through an MOU or other agreement helps to define the roles and responsibilities of liaisons, State DOTs, and resource agencies. When developing an agreement, State DOTs and resource agencies should consider:

- **Making the agreement renegotiable.** Developing renegotiable agreements can save time when existing agreements need to be updated to include additional (or fewer) positions, new funding structures, or other changes.

- **Modeling new funding agreements from existing agreements within the same State and/or other States,** if such agreements already exist. This can expedite the development of new agreements and allow agencies to better identify what new tasks or objectives to include.

7 Stage 5: Implementing and Managing the Program

This section addresses the final steps needed to implement and manage a program once a formal interagency agreement is in place, with a focus on navigating interagency relationships. The first step is to find and hire qualified liaisons and to determine how the State DOT should be involved in this hiring process. After the liaisons are hired, it is important to determine their work priorities clearly so that both the State DOT and the resource agency have the same expectations about their workload. One important factor to consider when determining work priorities is the level of involvement that liaisons should have in Section 6001 transportation planning activities.

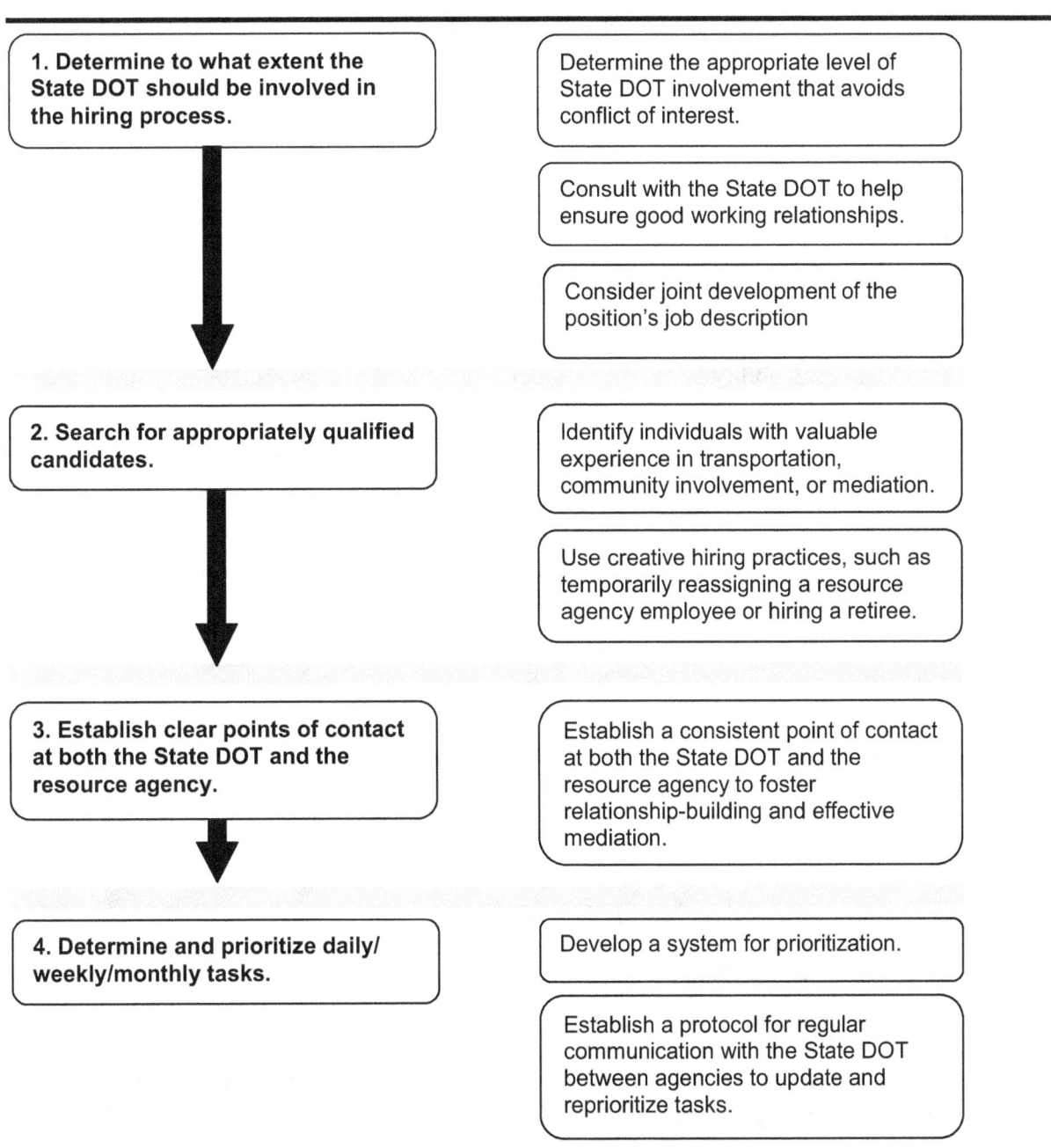

Figure 7. Implementing and Managing the Program

7.1 Finding and Hiring Liaisons

7.1.1 Essential Skills and Qualities

The success of the funded positions program is largely dependent upon finding the right person to staff the funded position. Both State DOT and resource agency staff noted the benefits of hiring liaisons who had prior experience in transportation project delivery or who had previously worked in regulatory or resource agencies.

Staffing the liaison position with experienced or senior-level employees often provided more abstract benefits beyond streamlining the permitting process, such as helping to initiate and improve processes or providing input on higher-level policy issues. For example, one liaison, a senior employee with previous resource agency experience, developed a regional general permit that significantly streamlined the project development process. This effort was not part of the liaison's required workload, yet as someone who interacted with both resource agency and State DOT staff the individual was able to identify areas of improvement and initiate solutions that benefited multiple parties.

To address the demands of the funded liaison position, State DOTs and resource agencies recognized the benefit of hiring individuals who demonstrated some of the following skills and qualities:

- Excellent written and oral communication skills.
- A willingness and ability to act as a "go-between" for two agencies that have different missions and/or organizational cultures.
- Experience in conflict mediation and a proven track record of problem-solving.
- An ability to strongly represent and articulate the resource agency's mission and to understand different viewpoints. This quality was deemed particularly important in situations where there was a need to build better working relationships between the State DOT and the resource agency due to past conflict or communication difficulties.
- A willingness to support the resource agency's goals and mission even in contentious situations.
- Background and experience in transportation planning to facilitate the liaison's work on State DOT projects.
- Motivation and self-direction.

7.1.2 Recruiting and Retaining Qualified Candidates

As previously described, shorter term lengths for liaison contracts often presented challenges in attracting qualified candidates, as many applicants did not want to accept positions without the assurance of permanent employment. To address this challenge, one State DOT increased the liaison pay rate, which led to some conflict at the host resource agency. Some State DOTs and resource agencies also faced budgetary constraints within their States, which led to additional funding constraints or hiring freezes.

Creative approaches that some State DOTs and resource agencies have used to hire qualified persons for the funded positions, especially when faced with difficult fiscal and hiring circumstances, are demonstrated in the following examples:

- In an attempt to overcome constraints on its ability to hire additional government FTEs, a resource agency hired a retired resource agency employee as a liaison through an existing contract with a consulting firm. Hiring the liaison as a consultant allowed the State DOT and the resource agency not only to work around the constraint on hiring new FTEs but also to acquire an individual with significant experience.

- In several cases, a liaison for a particular resource agency was hired from within the ranks of existing resource agency staff. Individuals who had previously worked as FTEs at the resource agency were able to apply their knowledge and experience to better serve both the resource agency and the State DOT.

- At one resource agency, an existing agency employee filled a liaison vacancy. That employee's original job became the termed position, while the liaison position became a career FTE position. The arrangement allowed the agency to fill the liaison position with a higher-grade employee who was already in a career position. The agency also ensured that the employee could return to the original position if funding for the liaison position were to suddenly be terminated.

- Several resource agencies filled their funded positions vacancies with individuals who had previously worked as liaisons in other States or resource agencies.

- In one case, a resource agency's recruitment process had hindered its ability to quickly hire a liaison to fulfill current workload demands, so the State DOT sent a staff member to work temporarily at the resource agency. Upon returning to the State DOT, the staff member brought an increased knowledge of the resource agency's practices.

- Several resource agencies made arrangements with the State DOT to receive funding for the equivalent labor-hours of one FTE. The resource agencies then divided the State DOT-focused workload between two or more positions while requesting funding for only one FTE. This arrangement provided the resource agencies with increased flexibility to assign tasks to multiple personnel, which also afforded the staff with some measure of job-task variety.

- One State DOT found additional State funding sources to hire consultants to supplement the work of existing liaisons at a particular resource agency. The resource agency welcomed the arrangement, as it did not add any administrative responsibilities and allowed utilization of the consultants to complete liaison-related work as necessary.

7.1.3 The Hiring Process

The involvement of State DOTs and resource agencies in the liaison hiring process varied throughout the agencies interviewed and often reflected broader relationships among them. In some cases, the State DOT contributed to making hiring decisions. In other cases, the resource agency was the sole decisionmaker on staffing.

The varied approaches that State DOTs and resource agencies took in the hiring process are demonstrated in the following examples:

- Some resource agencies expressed concern that State DOT involvement in the hiring process could be construed as "paying for permits." These agencies did not allow the State DOT to have any input in hiring decisions.

- One resource agency asked a State DOT to sit on the interview panel and provide input; however, the resource agency reserved the right to select the new hire.

- One State DOT worked collaboratively with the resource agency to develop job descriptions for several funded liaisons, while other job descriptions were developed solely by the resource agency.

- In another example of collaboration, a State DOT provided guidance on the resource agency's hiring process for one liaison, although the resource agency selected the new hire. The resource agency asked the State DOT whether it wanted to extend its involvement in the process, but the State DOT declined.

7.1.4 Setting Term Lengths

Most of the funded positions were created as termed positions, with the length of the contract term primarily driven by the number of years that the State DOT could commit to funding liaisons. Both State DOTs and resource agencies identified several challenges resulting from short-term positions, most notably difficulty in staffing and in retaining high-quality candidates. In addition, several resource agencies reported higher-than-average turnover rates for shorter-term liaisons, presumably due to staff leaving in search of more permanent positions. High liaison turnover not only placed additional administrative burdens on the resource agency but also hindered the relationship-building process between the State DOT and the resource agency. Furthermore, there was an increased risk of losing institutional knowledge when liaisons who had been trained or had developed familiarity with agency processes left the organization after a short period. On the other hand, some interviewees noted that having term positions allowed both the State DOT and the resource agency to "test" an individual's capabilities and ensure a good fit for the position.

One State experimented with different timeframes for the funded positions term and decided that five years was an ideal term length. A two-year term was too short to maintain staff continuity and prevent excessive turnover, while a term longer than five years led to difficulties in guaranteeing funding availability.

7.1.5 Program Management Structures

Some States with multiple districts or regions implemented centralized management structures whereby funded positions and management were co-located, while others implemented decentralized management structures whereby funded positions were dispersed. Both types of structure had advantages. For example, State DOTs and resource agencies with large programs or district offices that were physically located far apart found that decentralized, onsite management was more practical since it would be difficult for supervisors at one central office to oversee multiple liaisons' work at several different offices. A reported benefit of centralized management was the increased proximity of funded positions to their supervisors. For example, one liaison mentioned seeing the manager frequently due to the manager's location onsite at the office; this frequent contact enhanced work-task coordination. In addition, having a centralized management structure allowed some staff to focus on monitoring the terms of the agreement and others to concentrate on other aspects of the program, such as communicating work tasks to liaisons.

States with multiple districts developed methods to centralize funded positions management, including establishing regional coordinator positions to manage several liaisons. A reported benefit of a centralized management arrangement was that it encouraged more interliaison communication. Several different centralized management structures are illustrated in the following examples:

- In some cases, a regional coordinator was responsible for overseeing the terms of the funded positions agreement while an onsite field supervisor monitored the liaison's day-to-day work. Several funded positions and supervisors agreed that separating the administrative work from the project-based work helped to make the funded positions program more efficient.

- In one State, liaisons in a resource agency's central office worked on regional projects while remaining in contact with liaisons in the agency's field offices who worked on local projects.

Many State DOTs also reported having considered whether to develop one or several points of contact for funded liaisons at the State DOT. Some of the factors involved in this decisionmaking process were as follows:

- Liaisons, resource agencies, and State DOTs participating in larger funded positions programs preferred having a single, consistent State DOT point of contact. The presence of a State DOT staff member with comprehensive knowledge of the funded positions efforts helped to facilitate a good working relationship between the State DOT and the resource agency. Having a single point of contact also significantly helped to increase the level of trust and communication between the funded positions and the State DOT. A single point of contact could also function as a strong advocate for the program while helping to communicate work tasks to the liaisons.

- Smaller funded positions programs were often able to accommodate multiple State DOT points of contact. Such programs tended to have fewer bureaucratic issues to navigate and were conducive to increased interaction between funded positions and the State DOT even if there were multiple State DOT points of contact.

The key to successful management structures was the fostering of close working ties among all parties. Funded positions that worked closely with one another as well as with State DOT contacts generally felt they were able to better address the streamlining of goals than were those who worked in relative isolation.

7.2 Resolving Institutional/Interagency Relationship Issues

7.2.1 Determining Work Priorities

Determining work priorities is an important aspect of managing liaisons and ensuring that staff meet program goals and objectives. In setting these priorities, the extent of collaboration between the resource agency and the State DOT varied. In some cases, prioritization was a very collaborative process, while in other instances, the State DOT or resource agency set priorities with minimal input from others. It is critical to effectively communicate priorities; agencies should establish a process for communicating changed priorities to all parties, including the State DOT, the resource agency, and liaisons.

Examples of additional models for determining work priorities are as follows:

- In several cases, particularly in States that had established strong working relationships between the State DOT and the liaison, project prioritization occurred on an informal or ad hoc basis, with the liaison and State DOT staff communicating frequently and discussing projects and priorities as needed.

- In other cases, priorities were primarily determined through quarterly meetings between the State DOT and resource agency management as well as the funded liaison. During the meetings, participants reviewed current projects and identified upcoming projects. In addition, each agency discussed its views on project prioritization.

- In one case, the resource agency had primary responsibility for determining work prioritization, but it met with senior State DOT management on a quarterly basis to discuss policy issues and major challenges. The interagency meetings helped the resource agency to identify the State DOT's priority projects and to better manage time spent working on them.

- In another case, the funded position was responsible for performing project reviews for several State DOT district offices that often had competing priorities. The liaison worked individually with each district to identify priority projects, deadlines, and the severity of resource impacts, which allowed the individual to better manage the workload.

- One State DOT determined the priority of projects for the funded position. When priorities changed, the State DOT notified the liaison. The State DOT also reviewed the liaison's project "to do" list to ensure that the work was completed on the basis of the State DOT's priorities. Similarly, in another State, the State DOT headquarters office was responsible for coordinating among the district offices in order to present a comprehensive list of priorities to the resource agency.

- One resource agency initially based its work priorities on a "top-10" project list provided by the State DOT. It found that, over time, the State DOT was identifying most projects as top priorities. As a result, it decided to operate on a "first come, first served" basis, working on projects in the order in which the State DOT communicated them.

> *"We did not have any methods for letting the funded position know what the priority tasks were. This was a big lesson learned. We needed to have some ability to direct the funded position's work as part of the agreement. The first time was pretty loosey-goosey."*
> – State DOT
>
> *"[DOT] tells us what the priorities should be, and we take that advice seriously. We try to accommodate what they say and what the transportation industry needs as a priority. They know that better than us."*
> – Funded position

7.2.2 Mediating Conflict

Funded positions often had a role in mediating conflicts between State DOTs and resource agencies or between districts. When mediation was included as part of funded positions' responsibilities, most liaisons and State DOT employees found it helpful to establish procedures and levels of authority for conflict resolution. Some ways that agencies dealt with assigning mediation responsibilities were as follows:

- One resource agency assigned the funded positions to mediate conflict between districts. In this case, the State had a centralized management structure for funded positions that allowed them to interact frequently with their supervisors. This framework made it easier for liaisons to perform interdistrict mediation when needed.

- A resource agency developed a conflict resolution protocol for use in all joint State DOT/resource agency projects and dedicated three or four days per month for joint meetings if necessary.

- One resource agency made conflict resolution experience a significant factor in its hiring decisions for new funded positions.

7.3 Involving Liaisons in SAFETEA-LU Planning Activities

SAFETEA-LU Section 6001 (23 USC 134 and 135) made two significant changes to 23 USC Parts 133 and 134 that require a heightened consideration of environmental issues in the planning process for both metropolitan and statewide plans. The planning process must include a discussion of environmental mitigation activities and consultation with State, Tribal, and local agencies; such discussion must include an assessment of transportation plans alongside resource plans, maps, and inventories, if available. Some States chose to incorporate these activities into the funded positions program by involving funded liaisons in planning-related functions. FHWA's Planning and Environment Linkages (PEL) program provides tools and resources to support the implementation of these integrated planning activities, which are designed to lead to a seamless decisionmaking process that minimizes duplication of effort, promotes environmental stewardship, and reduces project delays.

In general, States choosing to support funded positions' work in the planning arena did so in several different ways, including:

- A State DOT designated, in an MOU with a resource agency, that funding could be used for specific planning activities. The same DOT decided to fund a one-year planning position at the resource agency. This DOT was the only agency interviewed that reported having a single position dedicated solely to SAFETEA-LU planning provisions. The DOT also reported that two additional funded positions spent some of their time on planning activities.

- In three States, funded positions were involved in planning to some or a limited extent, although no position was solely dedicated to planning activities.

- In one State, funded positions at some resource agencies were involved with planning activities while liaisons at other resource agencies were not.

- One liaison that had significant prior experience with initiatives incorporating planning and NEPA was specifically hired by the State DOT at a high-grade level to facilitate similar initiatives at the State level. The liaison was tasked to work on broader-level process initiatives to support project streamlining, such as developing a white paper to detail how the resource agency's priorities could best be communicated to Metropolitan Planning Organizations (MPOs).

Funded positions who worked on planning or planning-related activities gave examples of their direct and indirect involvement, which included:

- Reviewing the regional transportation and public participation plans of the State's MPOs.
- Developing long-range planning tools, including new permitting processes.
- Participating in corridor planning.
- Providing comments to MPOs on project alignment or meeting with MPOs to engage in early project coordination.
- Developing habitat connectivity maps at the request of the State DOT and resource agency to guide MPOs in future project planning.
- Developing white papers at the request of the resource agency to explore approaches for integrating planning and environmental review.
- Engaging in a statewide pilot project to develop an approach for regional environmental mitigation that will help to provide a framework for local planning agencies to address potential project impacts.
- Developing a web portal for local planning agencies, at the request of the State DOT, to provide an information clearinghouse related to planning. The portal includes information about critical species and conservation issues for MPOs to consider when developing long-range plans.
- Conducting general outreach to MPOs to learn more about planning processes.
- Contacting MPOs to learn which resources (e.g., maps) they might be able to share with the resource agency to facilitate consultation.

Funded positions who engaged in these types of planning-related tasks reported that their involvement had several short- and long-term benefits, such as more effective communication and coordination between the resource agency and the MPO. One liaison who worked to some extent in the planning arena reported that the work had helped local planning agencies to learn more about the resource agency's perspective on avoidance and mitigation issues. The liaison believed that this early coordination could help to address issues before they became problems later in project development. Two liaisons reported that their higher-level involvement in planning, such as developing white papers, had allowed the resource agency to more thoroughly consider how to interact with MPOs during project planning and to better convey the resource agency's objectives to MPOs.

Overall, liaisons' involvement in the planning arena facilitated communication, contributed to better working relationships between resource agencies and the State DOT, and helped to resolve potential red-flag issues or conflicts at an early stage of project development. Several liaisons anticipated that, as involvement in planning activities continued over time, the short-term benefits of better communication and coordination would likely lead to longer-term benefits of project streamlining.

Several State DOTs, resource agencies, and funded positions reported that, although liaisons had no current involvement in planning activities, they perceived a need to participate in them further in order to effectively accomplish environmental review duties, indicating that funded positions could help to support PEL-related goals.

Some State DOTs supported positions that would focus solely, or on a limited basis, on planning, while others wanted to incorporate this activity in the future. Other agencies expressed a reluctance to involve liaisons in planning. For example, one liaison believed that it would be beneficial for State DOT planning staff to increase their outreach to MPOs as a way to build stronger MPO–resource agency partnerships rather than involving the funded position. State DOTs reported that funded positions were directed to focus on project-oriented tasks rather than planning activities. This reluctance appeared to stem from a few specific concerns:

- Concern that involvement in planning might preclude the funded position from focusing on other tasks that were more directly related to the State DOT's mission or on more immediate environmental review and/or permitting requests.

- Differing missions between State and federal resource agencies. For example, one funded-positions manager indicated that State government should not focus on planning but rather on the environmental review process.

- Concern that planning activities were more open-ended than other project-related work and were difficult to specify or incorporate in the terms of a funding agreement.

- A perception that the funded position's involvement in planning-stage projects might create unrealistic expectations about the resource agency's or State DOT's comments on later approvals of a project.

- Difficulty in addressing the desire/need to involve the funded position in planning activities if these tasks were not specifically outlined in the funding agreement.

Discussions with State DOTs, funded positions, and resource agencies about involvement in planning-related tasks uncovered several common challenges and lessons learned:

- States that wished to involve funded positions in planning functions would benefit from having these tasks be well defined and from the process being thoroughly documented, for example, in the funding agreement.

- States should consider having one supervisor manage both the liaisons' planning and non-planning work. This could facilitate better communication between the liaison and resource agency management while making it easier for the resource agency to identify any benefits derived from the planning work.

- Liaisons who worked at the resource agency for longer periods often had opportunities to develop a strong rapport with the State DOT or local planning agencies. These relationships helped to lend credibility and trust to consultation processes.

- Liaisons with a previous background in planning can be valuable resources to assist with planning-related activities.

Recommendations: Implementing and Managing the Program

Finding and hiring liaisons: Agencies engaged in the hiring process should ensure that essential skills and qualities for a funded positions applicant are clearly defined. Some desirable qualifications include:

- Ability to act as a "go-between" for agencies with competing sets of demands.
- Strong diplomatic skills and ability to clearly articulate an agency's point of view in contentious situations.
- Experience in transportation planning, community development, and/or conflict resolution.

Resolving institutional/interagency relationship issues: Factors that contribute to positive interagency relationships Include:

- Developing interliaison networking opportunities even in decentralized programs. Funded positions who worked closely with one another often find it easier to meet program goals than do those who work in relative isolation.
- Clearly prioritizing State DOT projects for funded positions through quarterly meetings, regular phone contacts, or a "first come, first serve" policy. It is critical to effectively communicate priorities since they might change over time. Agencies should establish an arrangement for communicating changes in priorities to all parties involved.
- Ensuring that funded positions understand their role in mediating conflict. To facilitate mediation, strong working relationships among agencies must be developed and consideration given to creating formal procedures for how liaisons address conflict.

Involving liaisons in SAFETEA-LU planning activities: Involving funded positions in planning activities can lead to several long-term benefits, including facilitation of project streamlining and support of more integrated planning approaches. Reported shorter-term benefits included MPO familiarity with resource agency processes, better resource agency–MPO working relationships, and identification of initial red-flag issues that could later create conflict or delays in permitting. States choosing to incorporate planning activities into the funded positions program can do so in a number of ways, including:

- Funding one or several liaisons to work solely on planning activities.
- Encouraging liaisons to spend some portion of their time on planning tasks.
- Requesting liaisons to work on higher-level initiatives to address ways to improve resource agency–MPO coordination.

8 Stage 6: Evaluating Program Outcomes

In order to evaluate a funded positions program, it is important to consider the general benefits and challenges that funded positions provide. It is also crucial to design and institute both quantitative and qualitative performance measures that demonstrate the benefits that funded positions provide in streamlining project review and delivery.

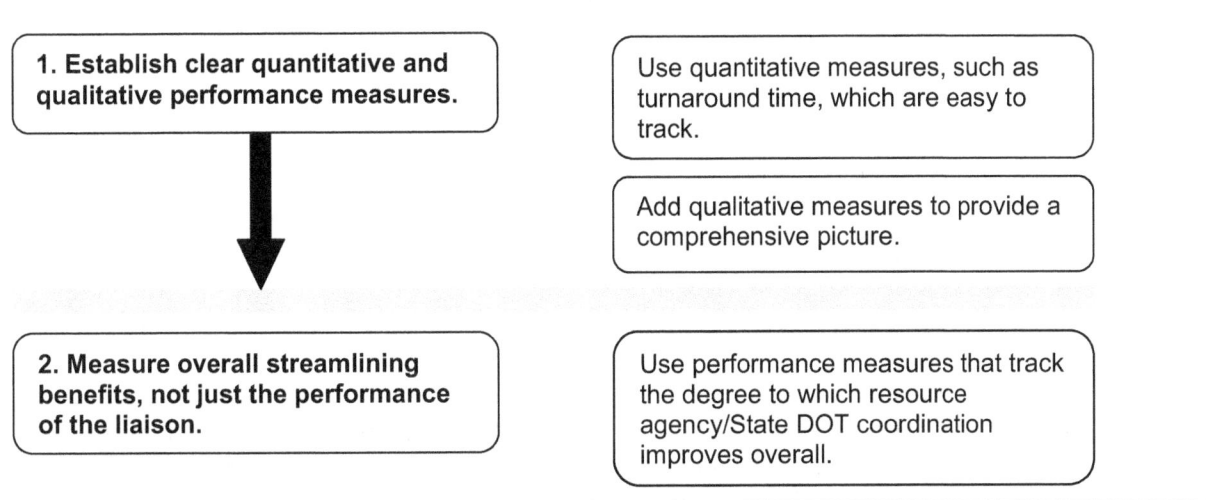

Figure 8. Evaluating Program Outcomes

8.1 Establishing Performance Measures

Most agencies have developed performance measures to evaluate whether the funded positions were meeting agency needs. For example, 80 percent of the interagency agreements reviewed for this report referenced performance measures as a component of the funded positions program. The following are common types of performance measures used in funded positions programs:

- According to a review in the *Transportation Research Record,* the most common accountability methods for new positions funded at other agencies are formal quarterly meetings.[20]

- State DOTs are tracking several output and intermediate outcome measures, including the number of days that environmental documents are at the offices of resource agencies and the amount of time spent on each project, on projects in general, and on nonproject administrative work, such as office meetings and required training.

- Where performance measures were specified, the primary metric was that a resource agency had to respond to the State DOT's submissions or complete project review within a 20-, 30-, or 45-day period. Some agreements specified calendar days, others mentioned working or business days. Other measures, such as those mentioned by the Idaho Transportation Department (ITD) and the Idaho State Historic Preservation Office (SHPO) agreement, were more qualitative; for example, "SHPO shall provide timely

---

[20]Marie Venner, "Measuring Environmental Performance at State Transportation Agencies." *Transportation Research Record*, Paper No. 03-4485, pp. 9–18.

review and findings for all cultural resource reports submitted by ITD as required by the National Historic Preservation Act of 1966."

- AASHTO found that State DOTs were collecting baseline data for some metrics. The metrics included permit turnaround times and the number of consultations that were being achieved prior to adding the funded position.[21]

- Most of the agreements briefly outlined how liaisons were to document their work or the number of hours of work completed. Liaisons were to keep daily or biweekly time records or timesheets and to provide quarterly status reports. In some instances, there would be monthly meetings, or project managers and agency directors might be expected to meet annually to evaluate the position and review program performance.

Only a few positions did not have any defined performance measures as part of the funded positions agreement. Without performance measures, these staff relied on ongoing communication with the State DOT to determine the agency's level of satisfaction with their work.

Most agencies that created performance measures reported having developed them as part of the funded positions agreements; many agencies included specific sections on performance objectives in the agreements. Some agencies also found it beneficial to conduct a self-assessment exercise to understand baseline needs. This exercise could initiate a process for developing performance measures by helping the agency to target critical areas in need of improvement, such as permit-turnaround times. The liaison could then be assessed on how well these needs were met. In other cases, performance measures were developed after the funded position was filled. However, only a few liaisons reported actually having been involved in this process.

In several cases where qualitative performance measures were developed, the resource agency and the State DOT worked collaboratively to outline appropriate measures. One State DOT met with each funded positions supervisor to discuss the supervisor's satisfaction with existing performance measures. The resource agency took the lead in suggesting additional measures to capture the full range of funded positions activities.

8.2 Quantitative versus Qualitative Performance Measures

The most commonly reported metrics used were quantitative. Such measures were typically based on timeframes or concurrence points outlined in the funding agreements. Common quantitative measures included:

- Project-review or permit-processing timelines.
- Number of meetings attended.
- Number of site visits.
- Amount of time spent per pay period on State DOT projects.
- Number of projects escalated to conflict resolution procedures.

---

[21] AASHTO, *op. cit.*, p. 27.

Most States reported that they had started their funded liaison program with just a few quantitative performance measures. Often, quantitative measures were derived from the terms of the funding agreement.

Once States gained experience with the use of funded positions, many agencies realized that there was a need to create qualitative measures to augment existing metrics. In some cases, resource agencies and funded positions advocated for the development of qualitative measures. Liaisons and their supervisors often found that the quantitative measures provided only a single, limited indicator of the funded positions' value and did not reflect the full picture of their contributions. For example, a resource agency noted that one of the State DOT's quantitative measures, tracking the number of consultations performed, did not capture the quality of those consultations. Several agencies working on establishing qualitative measures to accompany existing quantitative measures found this to be a difficult task. In one case, the resource agency reported that tracking timeframes and number of consultations was a relatively simple process but that developing metrics to measure the intangible benefits of funded positions was much harder. Intangible benefits included factors such as liaisons' contribution to stronger interagency relationships, quality of technical assistance, and enhanced environmental protection.

"We had a number of discussions with the DOT to get things more quantitative, but we came to the conclusion that we don't produce widgets. It's hard to come up with bean-counting measures of how much work you do; it's the qualitative part that is most important."
– Resource agency

"It's tough to measure the 'value added' that people can bring."
– Resource agency

Several staff in funded positions noted that, in order to be effective, both quantitative and qualitative performance measures should be based on outcomes or outputs over which they had control. Results that depend upon individuals or forces beyond the funded position should not be included as part of a performance assessment. For example, many liaisons commented that some activities, such as improving permit-turnaround times, were impacted by a variety of external factors and that it was unreasonable to assess their performance solely or primarily on the basis of these activities. Other factors largely beyond the liaisons' control included whether the State DOT submitted all of the necessary information to aid them in addressing an issue and whether there was a sudden change in priorities that affected timelines for completing a project.

8.3 Examples of Existing Performance Measures and Tools for Measurement

Examples of specific performance measures from existing funded positions programs are summarized below:

- Ohio DOT calculated specific time and cost benefits of their funded positions and found that the number of planned projects delivered within two weeks of their scheduled date increased from 70 percent in 1999 to 96 percent by 2007. In relation to the NEPA process, it reported cost savings of $30,000 to $100,000 as well as time savings of four to six months for large projects and two to three months for small projects.[22]

---

[22]NCHRP Peer Exchange Report, *op. cit.*, pp. 3-5–3-6.

- Agreements from Kentucky, Tennessee, and Texas include entire sections or attachments on performance objectives. For example, the TxDOT–TDEC agreement contains detailed performance goals and specified permit application review timelines: 30 days for processing general permits and issuing Notices of Coverage (NOCs), 90 days for processing Section 401 Certifications, and so on.
- The MDT–Montana FWP agreement provides a good example of measurable performance objectives. Specific language includes the following: "Supplemental staff will be responsible for initial identification and continued guidance and support through to successful completion, of at least five off-site wetland mitigation projects developed or programmed for development by MDT … in no less than three of the sixteen major Montana watersheds ... and shall result in wetlands credits for the benefit of highway construction projects."
- The Arkansas Highway Transportation Department (AHTD)–FHWA–USACE agreement details qualitative performance measures for each agency involved. All agencies were to jointly formulate recommendations to improve procedures and increase efficiency within three months of the time that the individual in the position to be funded by the agreement reports for duty; USACE was to ensure early coordination and prioritization of permit applications and significant improvement of existing processing times, inform AHTD if processing was expected to exceed normal times, and meet all timelines, and AHTD was to submit completed permit applications and consider altering permit applications as recommended.
- The PennDOT–USFWS agreement is the only one that references the need for baseline measures and requests that the liaison provide supporting data.
- The FDOT–Florida Department of Agriculture and Consumer Services (FDACS) agreement provides specific reporting requirements. Sample language includes: "FDACS shall review and respond to FDOT's ETDM review screens ... within 45 days of electronic notification that the project information has been uploaded …"; "… FDACS shall review and respond to FDOT submissions within 30 calendar days of receipt of complete project documentation"; and "FDACS will provide FDOT with quarterly status report to monitor program performance & verify utility of funded positions. Every 6 months, FDOT will issue a Performance Management report to discuss performance, efficiencies, timeframes, process issues and program activities."
- The TxDOT–TDEC agreement contains an entire section on performance objectives. This section also specifies permit-application- review timelines: 30 days for processing general permits and issuing NOCs, 90 days for processing individual ARAPs and Section 401 Certifications, and so on.
- The NCHRP Peer Exchange Report lists performance measures for 13 agencies.[23]

Table 9 lists tools that are useful for performance measurement and reporting.

---

[23] NCHRP Peer Exchange Report, *op. cit.*, p. 3-4.

Table 9. Performance Measurement and Reporting Tools*

Title (on Document)	Description	Contact (Website)
Initial Performance Standards – CDOT Position at USFWS	<ul><li>Describes performance standards for CDOT position at USFWS.</li><li>Performance standards include timeliness of documents and requests for concurrence, timeliness of biological opinions, availability for and attendance at meetings and field visits, phone responsiveness and customer service, substantive and constructive comments, quality of technical assistance, and documentation or self-evaluation.</li></ul>	http://environment.transportation.org/pdf/SDOT_funded/ColoradoDOT-FWSLiaisonPositionDescription.pdf
Colorado DOT–FWS Liaison 2003 Individual Performance Objectives	<ul><li>Describes performance objectives for CDOT/FWS Liaison.</li><li>Objectives include "review CDOT projects and documents for technical sufficiency," "work with Regions to avoid potential project delays," and "work closely with CDOT and other federal and state agencies in the development of programmatic agreements."</li></ul>	http://environment.transportation.org/pdf/DOT_funded/ColoradoDOT-InitialFWSLiaisonEvaluationQuestions.pdf
Performance Review Standards [for] Interagency Funding Agreements	<ul><li>Describes performance review standards for Interagency Funding Agreements.</li><li>Standards include providing document review, technical assistance (e.g., attendance at meetings, field visits, etc.), policy development and participation (e.g., involvement in departmental task forces), outreach and education (e.g., training development and presentations, district visits), and continuous quality improvement (e.g., attention to quality and streamlining at all stages of the interagency cooperative process).</li></ul>	http://environment.transportation.org/pdf/DOT_funded/PennDOT-PerformanceReviewStandards.pdf
Outline of USFWS–KYTC Monthly Reporting Template	<ul><li>Outlines monthly reporting template for USFWS–KYTC.</li><li>Template specifies cover letter, monthly report (to include summary of figures, contents of report, list of meetings, and summary of time spent on various items), and additional explanations/suggestions.</li></ul>	http://environment.transportation.org/pdf/DOT_funded/KYTCmonthlyReportingTemplate.pdf
[NCDOT's] Performance Assessment for (Agency) Positions Funded by State DOT	<ul><li>Serves as a performance assessment template for positions funded by State DOT.</li><li>Template includes fields for position title, division, summary of activities and accomplishments, and performance elements (e.g., core activities, meetings, policy/procedure, and planned goals/objectives).</li></ul>	http://environment.transportation.org/pdf/DOT_funded/NCDOTperformanceAssessmentForm.pdf
Letter from New Jersey Department of Environmental Protection, Natural and Historic Resources, Historic Preservation Office to New Jersey DOT, Bureau of Environmental Services and attached	<ul><li>Describes State Historic Preservation Office's (HPO) submittal of fourth-quarter Fiscal Year (FY) 2004 Summary of Accomplishments, as specified by an MOA with NJDOT; also requests final reimbursements for MOA services.</li><li>Attached Summary of Accomplishments details a schedule for HPO's review of FHWA and NJDOT submissions, summarizes other</li></ul>	http://environment.transportation.org/pdf/DOT_funded/NJDOT-SHPOQuarterlyPerformanceSummary.pdf

Title (on Document)	Description	Contact (Website)
Summary of Accomplishments	completed and outstanding work resulting from the HPO-NJDOT MOA, and provides documentation related to the MOA (e.g., HPO State DOT salary reimbursement receipt and New Jersey travel voucher).	
USFWS & NOAA Fisheries DRAFT Quarterly Reporting with WSDOT	• Describes format for quarterly agency reporting with WSDOT. • Quarterly agency reporting will occur every three months and will be combined with WSDOT liaison manager's comments to create a quarterly Liaison Performance Report. The reports will include a tabular section addressing measurable quantitative factors, such as number of biological assessments responded to, and a narrative section providing information about the effectiveness/quality of liaisons' work and accomplishments or any issues/problems encountered during the course of work.	http://environment.transportation.org/pdf/DOT_funded/WSDOT-FWS&NOAAFisheriesQuarterlyReporting.pdf

*DOT = Department of Transportation, USFWS = U.S. Fish and Wildlife Service, CDOT = Colorado DOT, KYTC = Kentucky Transportation Cabinet, NCDOT = North Carolina DOT, HPO = Historic Preservation Office, MOA = Memorandum of Agreement, FHWA = Federal Highway Administration, NOAA = National Oceanic and Atmospheric Administration, WSDOT = Washington State DOT.

Additional performance measurement tools that were not developed specifically for funded positions but are nonetheless relevant include:

- ***Measuring Environmental Performance at State Transportation Agencies*** (January 2007). This document discusses the purposes of performance measurement, indicators, and special challenges. Examples from States are also given. The report is available at http://trb.metapress.com/content/x61r686472g02m24/fulltext.pdf.

- ***WisDOT Performance Measures Library.*** This website is an index of performance measures — State, National, International; City, County, Regional, and Transit; Research and Resources; and Congestion and Traffic Operations. It can be found at http://www.wsDOT.wa.gov/Accountability/Publications/Library.htm.

- ***Measurement Initiatives – State Government.*** This website, prepared by a government consultant, links to state and NPO measurement initiatives. It can be found at http://www.seagov.org/initiatives/state_gov.shtml.

8.4 Use of Performance Measures

While agencies reported that the purpose of establishing performance measures was to evaluate whether funded positions addressed critical agency needs, only two State DOTs reported using the performance measures and quarterly reports during agreement renegotiations to determine the appropriate number of positions to support. Similarly, only two agencies reported using the results of the performance measures to assess whether changes to the funded positions program were warranted. When the funding positions contracts expired, one State reviewed the performance measures included in the annual reports and convened stakeholders to evaluate whether positions were achieving intended program goals. As a result of this review, the State terminated four positions. In other instances, States increased the number of positions because the review of performance measures indicated that the liaisons were providing significant benefits.

One State DOT identified the lack of baseline measures as the main impediment to using established measures to evaluate the funded positions program. The DOT reported that it was currently working to develop a database system to collect baseline data, which could then be used to compare current and past measures and to better evaluate the future benefits of the funded positions program.

One funded liaison noted that, in order to effectively assess the funded positions program, a meaningful programmatic assessment is required, and to understand the value of the program as a whole, it is insufficient to examine only individuals' performance. A programmatic assessment would involve an evaluation of both the resource agencies' and the State DOT's successes in meeting program goals. This type of assessment would also help to ensure that agencies are held accountable to standards.

8.5 Measurement of Streamlining of Benefits

To measure the streamlining of benefits, most agencies focused on the extent to which the liaison expedited or facilitated project permitting and provided quality customer service to the State DOT. Few agencies reported having developed metrics for evaluating the degree to which the program had enhanced overall agency cooperation, the amount of information-sharing that occurred between agencies, or the timeframe in which cooperative activities occurred.

To address this challenge, several liaisons identified a specific need for additional performance measures that evaluated the entire streamlining process. In one case, a liaison addressed this need by crafting a general, regional permit to improve streamlining for a State and then requesting that elements related to this effort be added to the liaison's performance appraisal.
Environmental streamlining and stewardship requires transportation agencies to work together with natural, cultural, and historic resource agencies to establish realistic timeframes for the environmental review of transportation projects. These agencies then need to work cooperatively to adhere to those timeframes while protecting and enhancing the environment. To achieve this goal, several agencies felt that it was important to provide increased decisionmaking power to higher-grade liaisons so that they could more easily enforce cooperative behaviors. Higher-grade employees may be better able to identify and understand multiple policy implications and could more frequently interact with senior-management structures. However, one State DOT mentioned that it had no recourse if resource agency employees at higher grades than the funded position did not stay on schedule. Project-level liaisons had little control over these scheduling delays and did not want to be assessed on this basis.

8.6 Administrative Reporting and Providing Feedback

Many State DOTs and resource agencies found that hourly timesheets create an undue administrative burden for the liaison who must fill them out and for the agency staff member who must review them. Creating less intensive or less frequent reporting methods can reduce the administrative burden. However, collecting information on the progress of projects and the time spent on activities is important; agencies should consider the appropriate reporting frequency that does not place undue administrative burden on the liaison or agency staff. When establishing reporting requirements for a funded position, State agencies should also consider the level of detail needed to provide accurate reporting.

Successful collaboration between State DOTs and resource agencies in a funded positions program usually occurred when the agencies jointly identified staff members to manage the terms of the funded positions agreement:

- One resource agency representative noted that having an agreement without oversight can be the same as not having an agreement at all. The agreement may "stall" without proper oversight.

- A funded positions supervisor reported that ensuring oversight over the agreement sends a message about the State DOT's and resource agency's level of commitment to the program.

Where there are clear management structures at the resource agency and the State DOT, quarterly reporting can be a valuable tool to assess the achievements of the funded position. Several resource agency and State DOT representatives mentioned that quarterly reports helped to provide a framework for examining progress and ensuring productive meetings between the State DOT and the resource agency.

State DOTs reported a variety of methods for collecting and providing feedback regarding the performance of the funded position. These methods included quarterly or annual status reports, interagency meetings to discuss performance and issues, or a combination of both.

Several State DOTs required that funded positions and resource agencies submit status reports on a quarterly or annual basis. Some State DOTs developed specific forms or templates, while others allowed the resource agency to determine what information was important to report to the State DOT. In general, the status reports included a section for collecting quantitative information as well as a section for the liaison to provide more open-ended feedback. While the use of status reports was common, several resource agency staff reported the limitations of tracking performance solely through reports. In one state, a funded position found that the State DOT's status report presented a narrow view of the staff's accomplishments. To provide a broader view of the activities performed during each reporting period, the liaison developed a narrative-based report to accompany the State DOT's spreadsheet.

"Billable hours are accumulated by the Human Resources staff person, and she forwards them to [State] DOT staff along with quarterly reports. I have never heard anything back from those reports. I don't even know who to go to, to talk about that."
– Funded position

"Part of the stress about my job is not knowing what [State] DOT thinks about it and whether it's helpful for them. I would like feedback from them."
– Funded position

Many liaisons, State DOTs, and resource agencies reported benefits of status reporting, such as keeping all parties up to date on the accomplishment of program goals and objectives. However, many liaisons commented that they rarely received any feedback from the State DOT after the status reports were submitted. In situations where the chain of command was not well established, two liaisons reported that they had submitted quarterly reports but had not received any response from the State DOT indicating that it had received or reviewed the reports. Some funded positions expressed frustration about the lack of State DOT feedback; one noted job-related stress that was due in part to not knowing how the

State DOT felt about the liaison's performance. Other liaisons who had not received any feedback from the State DOT operated on the premise that "no news is good news."

In place of or in addition to status reports, some State DOTs held regular meetings with the resource agency manager and/or the liaison to discuss performance. These meetings typically took place on a quarterly or annual basis and provided an opportunity for State DOTs, resource agencies, and liaisons to discuss and resolve issues. Face-to-face meetings also provided a means to provide feedback on some of the more intangible aspects of the funded position. In some cases, the funding agreement stipulated the occurrence of interagency meetings, yet such meetings took place less often than specified or sometimes not at all. One State noted that the physical distance between the State DOT and resource agency staff was one of the primary factors contributing to meeting infrequently.

8.7 Performance Appraisals

In addition to being evaluated by the State DOT, funded positions typically received an internal performance evaluation from resource agency management. Such reviews were part of the standard performance appraisal process conducted for all staff. In most cases, the funded position's internal performance objectives were based upon the measures developed as part of the performance measures outlined by the State DOT. Other resource agencies reported having separate metrics for evaluating staff positions as part of an internal review process.

The majority of resource agencies did not share outcomes of internal staff reviews with State DOTs. Several agencies noted that sharing these results was outside the realm of the funded positions agreements.

A recurring performance appraisal issue reported by liaisons and supervisors was the relative ease of developing and applying quantitative performance measures as compared with the relative difficulty of developing qualitative measures. Despite these reported difficulties, many interviewees believed that both quantitative and qualitative performance measures should be included in a performance appraisal.

Examples of qualitative performance measures that are applicable to funded positions include:

- The extent to which the liaison conducted outreach to the State DOT. Many State DOT officials indicated that the most effective liaisons were those who consistently contacted them to discuss priorities and provide regular project-status updates.

- The liaison's level of understanding regarding organizational processes and task priorities. One State DOT mentioned that the agency had eliminated a funded position because the liaison's lack of understanding of State DOT organizational procedures had produced inefficiencies and delays.

- The liaison's ability to build relationships and effectively communicate with State DOT and resource agency personnel. Several State DOTs mentioned the importance of hiring liaisons with strong communication skills and proven problem-solving abilities to help resolve issues in sometimes contentious situations.

- Successful navigation of the permitting process and an ability to negotiate between the State DOT and the resource agency. One State DOT reported that it was more likely to continue funding positions if the liaison could document practices that improved interagency relationships and expedited the permitting process.

- The quality of communications/comments provided to the State DOT. Several State DOTs stated that the quality of quarterly reports and other documents were critical factors highlighting the continued need for the funded positions program.

Recommendations: Establishing Performance Measures

Establishing performance measures: State DOTs and resource agencies can develop performance measures to assess whether the funded positions are meeting agency needs. Liaisons can also participate in creating performance measures. Effective performance measures should be based on outcomes or outputs over which the funded position has control.

Quantitative versus qualitative measures: Agencies and liaisons should consider developing both qualitative and quantitative measures to ensure a more robust and comprehensive picture of liaisons' contributions and accomplishments.

- Quantitative measures could include factors such as:
    - Extent to which the liaison met project-review or permit-processing timelines.
    - Number of site visits.
    - Amount of time spent per pay period on State DOT projects.
- Qualitative measures include the extent to which liaisons helped to foster strong working relationships and increased communication between the State DOT and the resource agency.

Conducting a baseline study: Conducting a self-assessment exercise to understand baseline needs is important when developing effective performance measures. Agencies need to understand the coordination process before hiring a liaison to determine whether the funded positions program has produced an improvement in coordination and project delivery.

Creating reporting requirements: To document progress made, States can establish reporting requirements for a funded position. Reports can be provided on an annual or quarterly basis or can take place during regularly scheduled resource agency/State DOT meetings. In addition to providing feedback on a liaison's performance, results of performance appraisals or reporting can be used to:

- Establish a continued need for the funded positions program.
- Document the specific number of positions needed.
- Compare current and past measures and better evaluate the future benefits of the funded positions program.

APPENDIX A: Interview List

The table below shows the interviews conducted for this project, including the date of the interview, the name and agency of the interview participant(s), and the participants' titles and/or offices (if known).

Table D-1. Information from Interviews That Were Conducted

State	Participant Name	Participant Title	Participant Agency*	Interview Date
California	Connell Dunning	Transportation Lead, Environmental Review Office	EPA, Region 9	7/28/08
	Susanne Glasgow	Deputy District Director	Caltrans, District 11	8/7/08
	Susan Sturges	Life Scientist	EPA, Region 9	8/15/08
	Shawna Pampinella	Acting Office Chief, Interagency Relations and Staff Development Office	Caltrans, Headquarters	8/25/08
	Kurt Roblek	Senior Biologist, Carlsbad Office	USFWS	8/28/08
	Roberta Gerson	Regional Transportation Coordinator	USFWS	10/29/08
	Douglas Hampton	Fisheries Biologist	NMFS	10/14/08
	Tami Grove	Statewide Liaison Coordinator	California Coastal Commission	10/24/08
	Dick Butler	Division Manager	NMFS	11/5/08 (joint interview)
	Scott Hill	Area Office Supervisor	NMFS	
Florida	Larry Barfield	Environmental Management Office	FDOT	9/11/08 (joint interview)
	Buddy Cunill	Manager, Environmental Quality Performance Section	FDOT	
	Terry Gilbert	Biological Scientist	FWCC	8/27/08
	Madolyn Dominy	Water Management Division	EPA, Region 4	8/18/08
	Andy Phillips	Regulatory Division, Permitting Section	USACE	9/10/08
	Scott Sanders	Leader, Habitat Conservation Scientific Services	FWCC	8/20/08
	Heinz Muller	Regional NEPA Coordinator	EPA, Region 4	8/19/08
	Bob Barron	Project Manager, Regulatory Division	USACE, Jacksonville District	9/19/08

State	Participant Name	Participant Title	Participant Agency*	Interview Date
North Carolina	Brian Wrenn	Supervisor of Transportation Permitting Unit, Environmental Program Supervisor III	NCDENR, Division of Water Quality	10/17/08
	Cathy Brittingham	Transportation Projects Coordinator – Central Office	NCDENR, Division of Coastal Management	11/04/08
	Brian Cole	Supervisor, Transportation Permitting Unit, Raleigh	USFWS	10/29/08
	Marella Buncick	Fish and Wildlife Biologist	USFWS	10/30/08
	Ehren Meister	Contract Administrator, Office of Environmental Quality	NCDOT	9/25/08
Ohio	Tim Hill	Administrator, Office of Environmental Services	Ohio DOT	10/8/08
	Karen Hallberg		USFWS	11/6/08
	Mark Epstein	Department Head	Ohio Historical Society	11/3/08
South Carolina	Randy Williamson	Environmental Engineer	SCDOT	4/14/09 (joint interview)
	Tim Hunter	Environmental Operations Manager	SCDOT	
	Steve Brumagin	Project Manager	USACE	4/3/09
Tennessee	Doug Delaney	Assistant Chief of Environment and Planning	TDOT	1/30/09
	Lee Barclay	Supervisor, Cookeville Field Office	USFWS	2/12/09
	Ben West	Biologist, NEPA Program Office	EPA	2/09/09
Utah	Jim McMinimee	Director of Project Development	UDOT	2/6/09 (joint interview)
	Rebecka Stromness	Environmental Program Director	UDOT	
Washington	Bob Thomas	Program Manager	WSDOT	12/11/08

*EPA = Environmental Protection Agency, Caltrans = California Department of Transportation (DOT), USFWS = U.S. Fish and Wildlife Service, NMFS = National Marine Fisheries Service, FDOT = Florida DOT, FWCC = Fish and Wildlife Conservation Commission, USACE = U.S. Army Corps of Engineers, NCDENR = North Carolina Department of Environment and Natural Resources, NCDOT = North Carolina DOT, SCDOT = South Carolina DOT, TDOT = Tennessee DOT, UDOT = Utah DOT, WSDOT = Washington State DOT.

APPENDIX B: Selected Bibliography

Reports and Articles
AASHTO Center for Environmental Excellence. *DOT-Funded Positions and Other Support to Resource and Regulatory Agencies, Tribes, and Non-Governmental Organizations for Environmental Stewardship and Streamlining Initiatives,* May 2005. Available at http://environment.transportation.org/pdf/DOT_funded/DOT_funded_positions_report.pdf.

Federal Highway Administration. Environmental Technical Assistance Program — Natural Resources Technical Assistance Program Alert: Number and Distribution of DOT-Funded Positions at Resource Agencies and Analysis of Trends Since 2001. August 2003. Available at http://environment.fhwa.DOT.gov/strmlng/etapreport.asp.

____. Interagency Funding Agreements Foster Streamlining: FHWA's Guidance on Use of TEA-21 Funds to Expedite Reviews. *Successes in Stewardship*, August 2002. Available at http://environment.fhwa.DOT.gov/strmlng/newsletters/aug02nl.asp.

Hendron, Patricia, with Wayne Kober. *Peer Exchange Series on State and Metropolitan Transportation Planning Issues Meeting 2: Work Assignments and Performance in External Positions Funded by State DOTs*. December 2007 draft (unpublished).

Venner Consulting. *State DOT Positions at Resource Agencies: Distribution, Limitations, Accomplishments, and Maintaining Accountability*. Prepared for AASHTO ETAP Standing Committee, NR-01-15, August 17, 2001. Available at http://nepa.fhwa.DOT.gov/ReNEPA/ReNepa.nsf/All+Documents/E595D1BE71F40E9985256BB2007F3637.

Venner, Marie. "Measuring Environmental Performance at State Transportation Agencies," *Transportation Research Record 1859*, Paper No. 03-4485, 2003.

Agreements, Letters, Memoraundums, Manuals, and Tools (by State)
Arkansas. Cooperative Agreement between United States Department of the Interior, Fish and Wildlife Service and United States Department of Transportation, Federal Highway Administration, and Arkansas Highway and Transportation Department. January 28, 2003.

____. Memorandum [between] Arkansas State Highway and Transportation Department [and] Arkansas Department of Heritage. October 9, 2000.

Arizona. Operating Agreement: The Integration Process Relative to the National Environmental Policy Act and Section 404 of the Clean Water Act. February 8, 2006.

____. Letter of Intent between the Arizona Department of Transportation and the Arizona State Historic Preservation Office Arizona State Parks. March 29, 2000.

____. Letter of Intent between the Arizona Department of Transportation and the U.S. Army Corps of Engineers. March 16, 2000.

California. Standard Agreement (STD 213) [between] Department of Transportation and U.S. Fish and Wildlife Service. December 1, 2004.

Colorado. Colorado DOT-FWS Liaison 2003 Individual Performance Objectives. n.d.

____. Initial Performance Standards – CDOT Position at USFWS. n.d.

Florida. Cunill, Buddy. Exhibit A-3 Amendment to NWFWMD 3-Year Funding Agreement. Florida Department of Transportation. July 16, 2007.

____. Exhibit A-1 Amendment to USFS 5-Year Agreement. Florida Department of Transportation. June 13, 2007.

____. Exhibit A-2 Amendment to NWFWMD 3-Year Funding Agreement. Florida Department of Transportation. January 9, 2006.

____. Exhibit A-1 Amendment to Northwest Florida Water Management District [NWFWMD] ETDM Funding Agreement. Florida Department of Transportation. August 15, 2005.

Florida. Appendix B: Federal Highway Administration and Federal Transit Administration Agency Operating Agreement. January 15, 2003.

____. Appendix C: US Army Corps of Engineers (USACE) Agency Operating Agreement. March 30, 2004.

____. Appendix D: National Marine Fisheries Service (NMFS) Agency Operating Agreement. February 9, 2004.

____. Appendix H: U.S. Fish and Wildlife Service (USFWS) Agency Operating Agreement (AOA). January 7, 2003.

____. Appendix I: Florida State Historic Preservation Officer (SHPO) and Advisory Council on Historic Preservation (ACHP) Agency Operating Agreement (AOA). August 15, 2003.

____. Appendix L: US Forest Service (USFS) Agency Operating Agreement. March 15, 2003.

____. Appendix M: Agency Operating Agreement [between] Florida Department of Environmental Protection (DEP) [and] Four Water Management Districts (WMDs). October 1, 2004.

____. Appendix N: Northwest Florida Water Management District (NWFWMD) Agency Operating Agreement (AOA). June 26, 2003.

____. Appendix O: Florida Fish and Wildlife Conservation Commission (FWCC) Agency Operating Agreement. November 12, 2002.

____. Appendix R: Florida Department of Community Affairs (FDCA) Agency Operating Agreement. August 25, 2003.

____. Appendix S: FDACS Agency Operating Agreement [with the Florida Department of Agriculture and Consumer Services]. October 7, 2003.

____. *FDOT ETDM Funded Positions Reference Manual.* January 2006.

____. Funding Agreement between State of Florida, Department of Transportation and United States Department of Transportation, Federal Highway Administration and South Florida Water Management District for Implementation of the Efficient Transportation Decision Making Process. February 1, 2007.

____. Funding Agreement between State of Florida, Department of Transportation and United States Department of Transportation, Federal Highway Administration and State of Florida, Department of Environmental Protection for Continuance of the Efficient Transportation Decision Making Process. January 1, 2007.

____. Funding Agreement between State of Florida, Department of Transportation and United States Department of Transportation, Federal Highway Administration and Southwest Florida Water Management District for Implementation of the Efficient Transportation Decision Making Process. October 1, 2006.

____. Funding Agreement between State of Florida, Department of Transportation and United States Department of Transportation, Federal Highway Administration and Suwannee River Water Management District for Implementation of the Efficient Transportation Decision Making Process. October 1, 2006.

____. Funding Agreement between Florida Department of Agriculture and Consumer Services (FDACS) and State of Florida, Florida Department of Transportation (FDOT) and United States Department of Transportation Federal Highway Administration (FHWA). July 1, 2006.

____. Funding Agreement between National Marine Fisheries Service (NMFS) and State of Florida, Florida Department of Transportation (FDOT) and United States Department of Transportation Federal Highway Administration (FHWA). July 1, 2006.

____. Funding Agreement between United States Department of Agriculture, Forest Service National Forests in Florida (USFS) and State of Florida, Florida Department of Transportation (FDOT) and United States Department of Transportation Federal Highway Administration (FHWA). July 1, 2006.

____. Funding Agreement between United States Environmental Protection Agency (USEPA) and State of Florida, Florida Department of Transportation (FDOT) and United States Department of Transportation Federal Highway Administration (FHWA). January 1, 2006.

____. Funding Agreement between Florida Department of Community Affairs (FDCA) and State of Florida, Florida Department of Transportation (FDOT) and United States Department of Transportation Federal Highway Administration (FHWA). December 19, 2005.

____. Funding Agreement between United States Department of the Interior Fish and Wildlife Service (USFWS) and State of Florida, Florida Department of Transportation (FDOT) and United States Department of Transportation Federal Highway Administration (FHWA). October 1, 2005.

____. Funding Agreement/Reimbursable Agreement between National Park Service (NPS) Agreement #501705001 and State of Florida, Florida Department of Transportation (FDOT) and United States Department of Transportation Federal Highway Administration (FHWA). July 1, 2005.

____. Funding Agreement between Northwest Florida Water Management District (NWFWMD) and State of Florida, Florida Department of Transportation (FDOT) and United States Department of Transportation Federal Highway Administration (FHWA). July 1, 2005.

____. Funding Agreement between Jacksonville District, US Army Corps of Engineers (USACOE) and State of Florida, Florida Department of Transportation (FDOT) and United States Department of Transportation Federal Highway Administration (FHWA). March 30, 2004.

____. Funding Agreement between Florida State Historic Preservation Officer (SHPO) and State of Florida, Florida Department of Transportation (FDOT) and United States Department of Transportation Federal Highway Administration (FHWA). August 15, 2003.

____. Funding Agreement between State of Florida, Department of Transportation and United States Department of Transportation, Federal Highway Administration and St. Johns River Water Management District for Implementation of the Efficient Transportation Decision Making Process. n.d.

____. Master Agreement between FDOT, FHWA, DEP, and Four WMDs: Implementing the Efficient Transportation Decision Making Process in Florida. October 1, 2004.

____. Master Agreement between FDOT, FHWA, DEP, and Four WMDs: Implementing the Efficient Transportation Decision Making Process in Florida [with the Florida Department of Environmental Protection]. October 1, 2004.

____. Master Agreement[:] Implementing the Efficient Transportation Decision Making Process in Florida [with the Southwest Florida Water Management District]. October 1, 2004.

____. Master Agreement[:] Implementing the Efficient Transportation Decision Making Process in Florida [with the Suwannee River Water Management District]. October 1, 2004.

____. Master Agreement[:] Implementing the Efficient Transportation Decision Making Process in Florida [with the U.S. Army Corps of Engineers]. March 30, 2004.

____. Master Agreement[:] Implementing the Efficient Transportation Decision Making Process in Florida [with the US Forest Service]. March 15, 2004.

____. Master Agreement[:] Implementing the Efficient Transportation Decision Making Process in Florida [with the National Marine Fisheries]. February 9, 2004.

____. Master Agreement[:] Implementing the Efficient Transportation Decision Making Process in Florida [with the Florida Fish and Wildlife Conservation Commission]. November 12, 2003.

____. Master Agreement[:] Implementing the Efficient Transportation Decision Making Process in Florida [with the Florida Department of Agriculture and Consumer Services]. October 7, 2003.

____. Master Agreement[:] Implementing the Efficient Transportation Decision Making Process in Florida [with the Florida Department of Community Affairs]. August 25, 2003.

____. Master Agreement[:] Implementing the Efficient Transportation Decision Making Process in Florida [with the State Historic Preservation Office]. August 15, 2003.

____. Master Agreement[:] Implementing the Efficient Transportation Decision Making Process in Florida [with the Northwest Florida Water Management District]. June 26, 2003.

____. Master Agreement[:] Implementing the Efficient Transportation Decision Making Process in Florida [with the Federal Highway Administration]. January 15, 2003.

____. Master Agreement[:] Implementing the Efficient Transportation Decision Making Process in Florida [with the U.S. Fish and Wildlife Service]. January 7, 2003.

____. Master Agreement[:] Implementing the Efficient Transportation Decision Making Process in Florida [with the Natural Resource Conservation Service]. October 21, 2002.

Idaho. Cooperative Agreement between the Idaho Transportation Department and the Idaho State Historic Preservation Office. March 8, 2002.

Indiana. Memorandum of Understanding [between] the Indiana Department of Transportation and the Indiana Department of Natural Resources. 2005.

Kentucky. Cooperative Agreement between the United States Fish and Wildlife Service, Federal Highway Administration, and the Kentucky Transportation Cabinet, Relative to Priority Project Review. July 1, 2004.

____. Outline of USFWS–KYTC Monthly Reporting Template. n.d.

Louisiana. Interagency Agreement Among United States Department of the Interior, U.S. Fish and Wildlife Service (USFWS), U.S. Department of Transportation Federal Highway Administration (FHWA), and the Louisiana Department of Transportation and Development (LA DOTD) State Project No. 737-99-0532 Federal Aid Project No. STP-9902(515) Regarding Reimbursement for Expedited Environmental Review and Enhanced Resource Agency Coordination Statewide. March 6, 2006.

____. Interagency Agreement Among United States Department of the Interior, U.S. Fish and Wildlife Service (USFWS), U.S. Department of Transportation Federal Highway Administration (FHWA), and the Louisiana Department of Transportation and Development (LA DOTD) Regarding Reimbursement for Expedited Environmental Review and Enhanced Resource Agency Coordination. 2002.

____. Interagency Agreement between the Louisiana Department of Transportation and Development (LA DOTD) and the Louisiana Department of Culture, Recreation and Tourism, Office of Cultural Development, Division of Archaeology (Division) Regarding Reimbursement for Expedited Environmental Review. 2003.

Maine. Maine Department of Transportation and Maine Department of Conservation Natural Areas Program Biennial Agreement for Services. 2004.

____. Maine Department of Transportation and Maine Historic Preservation Commission. Fiscal Year 2004 Agreement for Archaeological Services. 2003.

Mississippi. Memorandum of Agreement between United States Department of the Interior Fish and Wildlife Service (USF&WS) and Mississippi Transportation Commission (Commission) and, as a concurring party, United States Federal Highway Administration (FHWA). October 1, 2007.

Montana. Cooperative Agreement between the Montana Department of Fish, Wildlife and Parks (FWP), Federal Highway Administration (FHWA) and the Montana Department of Transportation Relative to Priority Highway Construction and Wetland Mitigation. June 2001.

____. Cooperative Agreement between the United States Department of the Interior Fish and Wildlife Service (USF&WS), Federal Highway Administration (FHWA), and the Montana Department of Transportation Relative to Priority Highway Construction Project Review. June 1999.

New Jersey. State of New Jersey Department of Environmental Protection [and] Natural and Historic Resources, Historic Preservation Office. Quarterly Performance Summary. August 11, 2004.

North Carolina. NCDOT Funded Position Program Reference Manual. November 2002.

____. Performance Assessment for (Insert Agency Name Here) Positions Funded by DOT. n.d.

Pennsylvania. Memorandum of Understanding [between] the Commonwealth of Pennsylvania, Department of Transportation and the Pennsylvania Department of Agriculture. 2004.

____. Memorandum of Understanding between the Commonwealth of Pennsylvania, Department of Transportation and the Pennsylvania Game Commission. MOU Number 430665. 2004.

____. Memorandum of Understanding [between] the Pennsylvania Department of Transportation and the Pennsylvania Historical and Museum Commission. 2004.

____. Intergovernmental Cooperation Act Agreement between United States Department of the Interior Fish and Wildlife Service ("USFWS") Commonwealth of Pennsylvania, Department of Transportation ("Department"). n.d.

____. Intergovernmental Cooperation Act Agreement between United States Environmental Protection Agency ("EPA") and Commonwealth of Pennsylvania, Department of Transportation ("Department"). n.d.

____. [Pennsylvania DOT] Performance Review Standards [for] Interagency Funding Agreements. n.d.

Tennessee. Agreement between the Tennessee Department of Transportation and the United States Department of Transportation Federal Highway Administration and the Tennessee Department of Environment and Conservation. November 16, 2007.

Texas. Interagency Agreement between the Federal Highway Administration and the U.S. Environmental Protection Agency Regarding Reimbursement for Expedited Environmental Review for Interstate Highway 69. October 29, 2003.

____. Interagency Agreement between the Federal Highway Administration and the U.S. Fish and Wildlife Service Regarding Reimbursement for Streamlined Environmental Review for Interstate Highway 69. October 29, 2003.

____. Cooperative Agreement between the Federal Highway Administration and the U.S. Army Corps of Engineers Regarding Funding for a Streamlined Environmental Review for Interstate Highway 69. October 7, 2003.

Washington. Washington State Department of Ecology Memorandum of Agreement with Washington State Department of Transportation. State Interagency Agreement/Agreement Number GCA 3841. November 2003.

____. Washington State Department of Transportation – US Fish and Wildlife Service & National Oceanic and Atmospheric Administration (NOAA) Quarterly Reporting [Information]. n.d.

APPENDIX C: Discussion Guides

State DOT Interview Guide

SAFETEA-LU 6002:
Assessing 6002(j) Funded Liaison Positions

Stage 1: Determining the Need for a Funded Position
During the first stage of the interview, we will try to uncover the decision-making processes behind a State DOT or resource agency's request for a funded position. Interviewees will be asked specific questions that get to the heart of how the need for a funded position was justified.

Describe the ways in which you evaluated the need for — and use of — 6002(j) funding.
- How are resource agencies/others involved in this evaluation?

Describe the activities you have used 6002(j)/139(j) funds for:
- Transportation planning activities that precede environmental review process
- Transportation project delivery activities, including consultation and permitting
- Training of agency personnel
- Information gathering and mapping
- Development of programmatic agreements
- Specific projects

Describe the process (if any) that was used to evaluate the need for a funded position and/or other types of support.

How were the priority tasks or work assignments for the funded position determined?

Did you face any issues or challenges in determining priority work assignments?
- Why do you think these issues/challenges existed?

What information would have been helpful to have when drafting the agreement?

What information or guidance (specific tools) do you think would be helpful in establishing your priorities for the funded position/activities?

Stage 2: Using Funded Positions
The second stage of the interview focuses on issues concerning the use and management of funded positions. Information collected will be used to develop lessons learned and best practices.

ADMINISTRATION
Can you describe the process for developing the funded position MOU/funding agreement?

- o Who was involved? Which agencies? Which staff? What level of staff/mgmt was involved?
- o At what stage did each agency get involved?
- o What was the impetus for their getting involved?
- o Was it a cooperative process? How long did the process take?
- o Did you look at examples from other states when you drafting the agreement?
- o Were there unique issues in your state that made it hard to find template agreements to follow?

Optional questions regarding interagency relationships:
- o Describe the working relationship between SDOT and the funded agency.
- o Have you outlined the responsibilities of all parties involved in the funded agreement?
- o How often do the agencies (SDOT, funded agency, FHWA) meet face to face?
- o How often do you talk with the agencies? Is there a formal process/schedule?
- o How do you balance resource agency needs/priorities/values and SDOT needs/priorities/values?
- o Are there any issues or challenges to creating a good working relationship and/or communication with the funded agency?
- o Do you think the relationship between the two agencies should be improved? If yes, how?
- o What information or guidance (specific tools) would be helpful in managing the interagency relationships?

Describe your view of the key terms of your MOU/Funding Agreement.
- o Were there any issues associated with administering the agreement?
- o If so, how do you think these can be diminished?

Has the MOU/Funding agreement been renegotiated? If so, can you describe the process for renegotiating the MOU/Funding agreement?
- o Who was involved? Which agencies? Which staff?
- o Was it a cooperative process?
- o How long did the process take?

Did you face issues or challenges in drafting/renegotiating the MOU, and why?

What information would have been helpful to have when drafting/renegotiating the agreement?

How are the finances/invoices managed for funded positions?
- o What information or guidance (specific tools) would be helpful in improving administration of the funded position and/or activities?

MANAGEMENT
Let's talk about management of the funded position. Can you tell me:

- Who manages the position?
- How are work tasks and program objectives communicated?
- What information or guidance would be helpful to better manage the funded position?

If the position is at the resource agency, do you have input into position management?

Are there any challenges to managing, hiring, and retaining funded positions?
- Why do you think these challenges exist?

Have you ever terminated a funded position arrangement? If so, please describe what happened.

TRAINING
Please describe the training process for funded positions.
- Do the funded positions receive training from the SDOT? If so, what does the training cover?
- Is the funded position expected to train the resource agency on SDOT practices/objectives?
- Is the funded position expected to train the SDOT on resource agency practices/objectives?

If position is at the resource agency, what input do you have on training decisions?

Can you identify any challenges and obstacles involved in the provision of training?

Why do you think these challenges exist?

PERFORMANCE MEASURES AND ASSESSMENT TOOLS
Is there a system or process in place to evaluate the performance of the funded position? If no, why not?

If so, how did you set up performance measures/tools?
- Who (what agencies) was/were involved in the process?
- If position is at the resource agency, how much input do you have in establishing performance measures and evaluations?

Can you describe some of the performance measures/assessment tools for the SDOT and the funded position (i.e., quantitative/qualitative)? Can you describe how you assess performance against the qualitative measures?

Have the performance measures helped you to assess the benefits of or improve the focus of the funded position?

Have the performance measures changed over time?
- If not, why not?

Are there any difficulties involved with defining and evaluating performance measures?

How have you overcome those difficulties?
- What information/tools have been (or would be) helpful to you in defining and or tracking performance measures?

PERFORMANCE REVIEWS AND REPORTING

Describe the performance review process or system used to assess positions (if at resource agency).
- How often do the reviews take place?
- Is there an appraisal of services provided, or an evaluation of the funded position's performance?

If the position is at the resource agency, how much input do you have in the performance review process/system?

PLANNING QUESTIONS

Are the positions involved in planning activities?
- For example, have you used the funded position to address the integrated planning and consultation requirements (Section 6001)?

Are there any obstacles to utilizing the funded position in planning activities?
- Why do you think these obstacles exist?

LESSONS LEARNED

Can you identify any lessons learned from your experience with funded positions?
- Could you provide some examples of 'dos' and 'don'ts?'

Can you identify any lessons learned from your experience with the resource agency with regard to the funded position(s)?
- Could you provide some examples of 'dos' and 'don'ts?'

Funded Position Interview Guide

SAFETEA-LU 6002:
Assessing 6002(j) Funded Liaison Positions

BACKGROUND (RESPONSIBILITIES, EXPECTATIONS, ADMINISTRATION)
With which of the following activities are you involved:
- __ Transportation planning activities that precede the environmental review process
- __ Transportation project delivery activities, including consultation & permitting
- __ Training of agency personnel
- __ Information gathering and mapping
- __ Specific projects

Briefly describe your job responsibilities.
- o How often do you attend public meetings?
- o Do you visit project sites or attend project level meetings at the SDOT?
- o Are you expected to address emergency permitting?
- o Do you act as a point of contact for FHWA or liaisons from other agencies?
- o Do you initiate or complete programmatic agreements?

How was the position's roles and responsibilities communicated to you, and by whom (SDOT, resource agency, or both)? Was a consistent set of roles and responsibilities communicated?

What is the level of communication between the agency you are housed in, and the agency whose work you are intended to accomplish? Is there a formal process/schedule in place? If so, do you have input into the schedule/agenda?

Describe your understanding of what the SDOT expects of your position. Also describe your understanding of what the resource agency expects of your position.

Describe your understanding of how the resource agency mission relates to the SDOT mission.

What is your perception of the different agency cultures?

Are you tasked with coordination activities or managing the interagency relationship? Do you think the relationship between the two agencies should be improved?

What information or guidance (specific tools) would be helpful in managing the interagency relationships?

Can you describe any issues or challenges that are related to general administration of your position? (Dealing with two agencies? Reporting? Finances/invoices? Performance Reviews? Performance Measures?) [Address in later categories.]

MANAGEMENT

Let's talk about the management of your position. Can you tell me:
- Who manages you? Which agency do you report to?
- Describe how your priority areas/work tasks/daily activities are determined. How are/were your work tasks and program objectives communicated to you? (How does the SDOT communicate its priorities to you? How does the resource agency communicate its priorities to you?)
- Is defining work tasks a cooperative process — did you have input into defining these tasks and objectives? (Or are the decisions made by the SDOT, Division Office, or Resource Agency)
- How do you balance resource agency needs/priorities/values and SDOT needs/priorities/values?

Can you identify any other management issues or challenges?

Why do you think these issues or challenges exist?

If issues or challenges exist, how do you think these can be addressed?

What information or guidance (specific tools) would be helpful in improving these management processes?

TRAINING

What training (both transportation and resource related) have you received? Did you receive training from the SDOT, or from the resource agency? If so, what does the training cover?

Do you feel you have received adequate training to conduct your job? If not, what areas do you think you need more training in?

Are you responsible for training SDOT or FHWA staff? If yes, describe what training you conducted and what the objectives were.

Are you responsible for training resource agency staff? If yes, describe what training you conducted and what the objectives were.

What are some of the challenges and obstacles involved in receiving/providing training? How do you think these can be addressed?

What information or guidance (specific tools) would be helpful in improving these training processes?

PERFORMANCE MEASURES AND ASSESSMENT TOOLS

Earlier, you identified your activities as …. are there performance measures that have been put in place to help you reach goals/outcomes related to these activities?
- If yes, please describe what the performance measures are (e.g., qualitative, quantitative, describe exact type).

- If there are no performance measures, do you know why not?

Do you know how those performance measures established? Who (what agencies) was/were involved in the process? Did you have any input in the process, or in your own performance plan?

Have the performance measures changed over time?

How often do you meet with SDOT staff/Resource Agency staff to discuss progress with performance measures?

Do you understand these performance measures? Do you think the performance measurements have been useful in terms of establishing guidance for your job or reaching certain goals/outcomes?

Is there a relative timeline comparison of when the task might have been completed if there was not a funded position?

Do you think the performance measures provide a fair evaluation/assessment of your work/agency performance?

Are there difficulties involved with defining and tracking performance measures? Describe.

How have you overcome those difficulties? Where do you go if you have a problem with meeting you performance objectives/measures?
- Do you have an opportunity to interact with other liaisons and compare working conditions or performance measures?

What information/tools have been (or would be) helpful to you in defining and or evaluating performance measures?

PERFORMANCE REVIEWS AND REPORTING
Describe the performance review process or the system used to assess your performance, if these processes/systems exist.
- How often do the reviews take place? Who/which agency is responsible for the reviews?
- Is there an appraisal of services provided, or an evaluation of your performance?

Do you know how/if the results of the reviews are used?

LESSONS LEARNED
Tell me about some of the lessons learned from your experience as a funded liaison for the SDOT and/or resource agency.
- Please provide some examples of 'dos' and 'don'ts.'

Resource Agency Interview Guide

SAFETEA-LU 6002:
Assessing 6002(j) Funded Liaison Positions

Stage 1: Determining the Need for a Funded Position

During the first stage of the interview, we will try to uncover the decision-making processes behind a SDOT or resource agency's request for a funded position. Interviewees will be asked specific questions that get to the heart of how the need for a funded position was justified.

Describe the ways in which you evaluated the need for — and use of — 6002(j) funding.
- How are SDOTs/others involved in this evaluation?

Describe the activities you have used 6002(j) funding for:
- Transportation planning activities that precede environmental review process
- Transportation project delivery activities, including consultation and permitting
- Training of agency personnel
- Information gathering and mapping
- Development of programmatic agreements
- Specific projects

Describe the process (if any) that was used to evaluate the need for a funded position and/or other types of support.

How were the priority tasks or work assignments for the funded position determined?

Did you face any issues or challenges in determining priority work assignments?

Why do you think these issues existed?
- What information would have been helpful to have when drafting the agreement?

What information or guidance (specific tools) do you think would be helpful in establishing your priorities for the funded position/activities?

Stage 2: Using Funded Positions

The second stage of the interview focuses on issues concerning the use and management of funded positions. Information collected will be used to develop lessons learned and best practices.

ADMINISTRATION

Can you describe the process for developing the funded position MOU/Funding agreement?
- Who was involved? Which agencies? Which staff?
- At what stage did each agency get involved?
- What was the impetus for their getting involved?

- o Was it a cooperative process? How long did the process take?
- o Did you look at examples from other states when you drafted the agreement?

Optional questions regarding interagency relationships:
- o Describe the working relationship between your agency and SDOT.
- o Are the responsibilities of all parties involved in the funded agreement clearly outlined?
- o How often do the agencies (SDOT, funded agency, FHWA) meet face to face? Under what circumstances? Regularly?
- o Is there a formal process/schedule?
- o How do you balance resource agency needs/priorities/values and SDOT needs/priorities/values?
- o Are there any issues or challenges to creating a good working relationship and/or communication with the SDOT?
- o Do you think the relationship between the two agencies should be improved? If yes, how?
- o What information or guidance (specific tools) would be helpful in managing the interagency relationships?

Describe your view of the key terms of your MOU/Funding Agreement.
- o Are there any issues associated with administering the agreement?
- o If so, how do you think these can be diminished?

Has the MOU/Funding agreement been renegotiated? If so, can you describe the process for renegotiating?
- o Who was involved? Which agencies? Which staff?
- o Was it a cooperative process?
- o How long did the process take?

Did you face issues or challenges in drafting/renegotiating the MOU, and why?

What information would have been helpful to have when drafting/renegotiating the agreement?

How are the finances/invoices managed for funded positions?
- o What information or guidance (specific tools) would be helpful in improving administration of the funded position and/or activities?

MANAGEMENT
Let's talk about management of the funded position. Can you tell me:
- o Who manages the position?
- o How are work tasks and program objectives communicated?
- o What information or guidance would be helpful to better manage the funded position?

If position is at the SDOT, do you have input into position management?

Are there any challenges to managing, hiring, and retaining funded positions?
Why do you think these challenges exist?

Have you ever terminated a funded position arrangement? If so, please describe what happened.

TRAINING
Please describe the training process for funded positions.
- o Do the funded positions receive training from the SDOT? If so, what does the training cover?
- o Is the funded position expected to train the resource agency on SDOT practices/objectives?
- o Is the funded position expected to train the SDOT on resource agency practices/objectives?

If position is at the SDOT, what input do you have on training decisions?

Can you identify any challenges and obstacles involved in the provision of training?

Why do you think these challenges exist?

PERFORMANCE MEASURES AND ASSESSMENT TOOLS
- o Is there a system or process in place to evaluate the performance of the funded position? If no, why not?

If so, how did you set up performance measures/tools?
- o Who (what agencies) was/were involved in the process?
- o If position is at the SDOT, how much input do you have in establishing performance measures and evaluations?

Can you describe some of the performance measures/assessment tools for the SDOT and the funded position (i.e., quantitative/qualitative)? Can you describe how you assess performance against the qualitative measures?

Have the performance measurements helped you assess the benefits or improve the focus of the funded position?

Have the performance measures changed over time?
- o If not, why not?

Are there any difficulties involved with defining and evaluating performance?

How have you overcome those difficulties?
- o What information/tools would be helpful to you in defining and or tracking performance measures?

PERFORMANCE REVIEWS AND REPORTING
Describe the performance review process or system used to assess positions (if at resource agency).
- How often do the reviews take place?
- Is there an appraisal of services provided, or an evaluation of the funded position's performance?

If the position is at the SDOT, how much input do you have in the performance review process/system?

PLANNING QUESTIONS
Are the positions involved in planning activities and how have these positions been involved?
- For example, have you used the funded position to address the integrated planning and consultation requirements (Section 6001)?

Are there any obstacles to utilizing the funded position in planning activities?
- Why do you think these obstacles exist?

LESSONS LEARNED
Can you identify any lessons learned from your experience with funded positions?
- Could you provide some examples of 'dos' and 'don'ts'?

Can you identify any lessons learned from your experience with the SDOT with regards to the funded position(s)?
- Could you provide some examples of 'dos' and 'don'ts'?

APPENDIX D: Funded Positions Agreements

Multiagency Agreements

– FDOT and FHWA Funding Agreement:
http://environment.transportation.org/pdf/DOT_funded/FDOT-FHWA_Agency_Funding_Agreement_All.pdf

– FDOT and FHWA Operating Agreement:
http://environment.transportation.org/pdf/DOT_funded/FDOT-FHWA_Agency_Operating_Agreement_All.pdf

Agreements with Federal and State Resource Agencies

– Caltrans–FWS Funding Agreement:
http://environment.transportation.org/pdf/DOT_funded/Caltrans-FWSAgreement.pdf

– WSDOT – WS Department of Ecology Memo of Agreement:
http://environment.transportation.org/pdf/DOT_funded/WSDOT-DepartmentofEcologyMOU.pdf

– FDOT–FWS Funding Agreement:
http://environment.transportation.org/pdf/DOT_funded/FDOT-FHWA-FWS_Funding_Agreement.pdf

– FDOT–FWS Operating Agreement:
http://environment.transportation.org/pdf/DOT_funded/FDOT-FHWA-FWSOperatingAgreement.pdf

– FDOT–SHPO Funding Agreement:
http://environment.transportation.org/pdf/DOT_funded/FDOT-SHPOFundingAgreement.pdf

– FSDOT–SHPO Operating Agreement:
http://environment.transportation.org/pdf/DOT_funded/FDOT-SHPOOperatingAgreement.pdf

– FDOT–FL Department of Environmental Protection Funding and Operating Agreements:
http://environment.transportation.org/pdf/DOT_funded/FDOT-FHWA-DEPWaterFundingOperatingAgreements.pdf

– FDOT–NW FL Water Management District Funding Agreement:
http://environment.transportation.org/pdf/DOT_funded/FDOT-FHWA-NWFWMD_Funding_Agreement.pdf

– FDOT–NW FL Water Management District Operating Agreement:
http://environment.transportation.org/pdf/DOT_funded/FDOT-FHWA-NWFWMD_Operating_Agreement.pdf

Programmatic Tools/Products Produced by Funded Positions:

See WSDOT's TPEAC product listed in Appendix C at
http://environment.transportation.org/pdf/DOT_funded/DOT_fpr_appendix_c.pdf.

Program Manuals and Guides

FDOT's Funded Position Reference Guide:
http://environment.transportation.org/pdf/DOT_funded/FloridaDOTFundedPositionsReferenceGuide.pdf

NCDOT's Funded Positions Program Manual:
http://environment.transportation.org/pdf/DOT_funded/NCDOTFundedPositionProgramManual.pdf

NCDOT's Performance Assessment Form:
http://environment.transportation.org/pdf/DOT_funded/NCDOTperformanceAssessmentForm.pdf

NCDOT's Ethics Policy:
http://environment.transportation.org/pdf/DOT_funded/NCDOTethicsPolicy.pdf

Workload Tracking and Performance Measurement Materials

ODOT's Monthly Focus Report for Environmental Services:
http://environment.transportation.org/pdf/DOT_funded/OhioDOTMonthlyFocusReportforEnvironmentalServices.pdf

WSDOT-FWS and NOAA Fisheries Quarterly Reporting Information—Quantitative and Narrative Information:
http://environment.transportation.org/pdf/DOT_funded/WSDOT-FWS&NOAAFisheriesQuarterlyReporting.pdf

Annual Report Tools and Formats

NCDOT's Annual Report and Performance Assessment Format:
http://environment.transportation.org/pdf/DOT_funded/NCDOTAnnualReportPerformanceAssessment.pdf

www.ingramcontent.com/pod-product-compliance
Lightning Source LLC
Chambersburg PA
CBHW081839170526
45167CB00007B/2845